Mechanics of Wood Machining

Etele Csanády · Endre Magoss

Mechanics of Wood Machining

Third Edition

Revised and Expanded

 Springer

Etele Csanády
Department of Wood Engineering
University of Sopron
Sopron, Hungary

Endre Magoss
Department of Wood Engineering
University of Sopron
Sopron, Hungary

ISBN 978-3-030-51483-9   ISBN 978-3-030-51481-5 (eBook)
https://doi.org/10.1007/978-3-030-51481-5

First edition published by the Department of Wood Engineering, University of West Hungary, Sopron, 2011
2nd edition: © Springer-Verlag Berlin Heidelberg 2013
3rd edition: © The Editor(s) (if applicable) and The Author(s), under exclusive license to Springer Nature Switzerland AG 2020

This Springer imprint is published by the registered company Springer Nature Switzerland AG
The registered company address is: Gewerbestrasse 11, 6330 Cham, Switzerland

Foreword

Wood is one of the most precious materials of the mankind. Since ancient times, wood material has been widely used and most of the utensils, shelters and dwellings were made of wood. Wood may have exceptional beauty in all shades of colour, and many species have excellent durability withstanding many hundreds of years. The oldest wooden buildings have withstood storms and disasters over 1300 years. At the same time, wood materials require one of the lowest energy inputs on the earth.

Since ancient times, the wood has been shaped by mechanical woodworking operations as we do today. We stand with astonishment and true admiration before a 20-m-high wooden column, somewhere in China or Japan, worked by hand with incredible accuracy—a thousand years ago!

Today, we have modern woodworking machines and hand tools; however, the quality of the raw material is continuously declining. We are forced to use this precious material with care and more economically, reducing waste as much as possible. To fulfil this requirement, efforts are being made worldwide to develop new machines, tools and technologies, for instance thin-kerf saws, wear-resistant tools, quality assessment, using of smaller size raw materials, etc.

The first comprehensive textbook on *Theory of Wood Processing* in Hungarian was published in 1994 and used by students at the Faculty of Wood Science, at the University of Sopron. This textbook contains a special chapter on the "Theory of Wood Cutting" and its related topics. Since the preparation of this book, almost 20 years have passed, and during this time new theoretical and practical knowledge accumulated and some new methods have been developed.

This book is the improved and expanded version of the above-mentioned chapter, supplemented with several related topics on vibration and stability of woodworking tools, the phenomenon of washboarding, tensioning methods and pneumatic clamping. The last chapter contains an important topic on wood surface roughness. In the last 15 years, fundamental research work has been done in this field and new approaches were developed. The new methods developed made it possible to characterize the different wood species with numerical values and to find general relationships concerning very different wood species.

It should be stressed that important results and new ideas have been published and, with all probability, will be published in many languages all over the world. Therefore, it is of great importance to consider all significant works done whenever and everywhere.

I hope this new textbook on the *Mechanics of Wood Machining* will contribute to the better understanding of physical phenomena associated with real woodworking processes. The material presented is designed for graduate and postgraduate students, and researchers and also to help practising engineers find the pertinent knowledge on this subject.

Sopron, Hungary   György Sitkei
February 2012   Professor Emeritus
                                                                                         Member of HAS

Preface to the Third Edition

This revised and expanded edition, six years after the appearance of the second edition, gave us the opportunity to rethink the whole material and to modify and complete it with new results now being developed. Reviews have been made in research areas in which considerable new materials and understanding have accumulated in the past six years. New updated materials are added in many sections. New Chapter 9 deals with the optimization of wood machining processes and discusses the latest developments in the field. Most of the new material is based in part on the authors' own research. About a third of book has been rewritten, and about a quarter of the figures are new. Several figures are modified for better understanding.

The book retains the general approach of the previous edition, and we have retained text and figures that are still relevant. Each topic in this book is developed from fundamental concepts as far as it was possible. This approach ensures that the developed relationships are less prone to obsolete by the time.

Some of the main revisions are

- Chapter 9 is newly added and describes the optimization procedure for wood machining process. The method is based on a newly developed engineering approach and abundantly illustrated with worked examples.
- Chapter 8 contains a much expanded treatment of surface roughness with many new materials resulted in the past six years.
- New Sect. 8.5.3 deals with the much debated question on the possible separation of roughness components due to machining and wood anatomy. Furthermore, Sect. 8.9 deals with the influence of wetting on the surface roughness.
- In Chap. 1, the relationships of chip deformation are extended to plastic flow, and Sect. 1.6 is newly added to describe the main regularities of hammer mills for grinding wood.
- In Chap. 2, for better characterization of thermal loads in tools, a detailed analysis of temperature gradients is included.

- In Chap. 4, new Sect. 4.6 deals with the energy requirement of sanding while new Sect. 4.7 describes generalized relationships for energy requirement calculations.

We hope this revised and expanded edition will continue to serve a need for all those working in the field of wood machining.

We acknowledge the important assistance of Dr. Z. Kocsis who prepared the new figures for publication. We are highly indebted to Prof. G. Sitkei for his continuous interest and many useful suggestions for improvements during the preparation of this new edition. We are also sincerely grateful to the staff of Springer-Verlag for their excellent co-operation.

Sopron, Hungary   Etele Csanády
2020   Endre Magoss

Preface to the Previous Edition

The importance of economical processing and manufacturing of various wood products is ever-increasing. During technological processes, wood materials may be exposed to various mechanical effects, e.g. the interactions between wood materials and different kinds of tools, and the general laws governing these interactions should be known.

The behaviour of wood materials deviates essentially from that of the generally known elastic materials. Wood has a strong anisotropy and viscoelastic properties which have highly aggravated the derivation of generally valid relationships. In the past time, because of our lack of understanding, knowledge of these processes was essentially restricted to empirical relationships and practical recommendations.

This textbook has attempted to develop a mechanics of wood machining in depth and extent so that it defines and delineates an area of knowledge that may be a new discipline in future. Each topic in this book is developed from fundamental concepts of mechanics. Where possible, mathematical treatment has been developed in quantitative terms to express specific wood–machine relations on the assumption that these relations should obey fundamental laws. Such a mechanics is capable of describing and predicting the interaction of wood and machine in terms of performance, quality and energy consumption. In many cases, working diagrams are elaborated to facilitate the consideration of several variables influencing the specific force and energy requirement of surface quality.

This textbook is an attempt to convey both definitive and practical information concerning wood machining to persons interested in various problems of wood machining. The topics were treated with the view in mind of presenting as much physical information as possible without excessive mathematical manipulation.

A distinct strength of this book is that it draws on truly international literature and not only English language research publications. At the same time, considerable part of this book relies on own research works.

The authors are especially indebted to Professor G. Sitkei for reading the completed manuscript and offering many useful suggestions.

The authors acknowledge the important assistance of Veronika Csanády and Lajos Reisz in the preparation of the manuscript for publication. They are also sincerely grateful to the staff of Springer-Verlag for their excellent work done with patience.

March 2012   Etele Csanády
                                                                 Endre Magoss

Contents

Chapter 1
Mechanics of the Cutting Process

1.1 Introduction

Machining plays a basic role in most procedures of the conversion of timber. Machining usually alters the shape, size and surface quality of wood. Machining occurs by cutting in most cases, and chips are the by-product. Some chips are used for the production of chipboard and some are used to generate energy. The chip may be the product in certain cases, and then we talk about technological or target chips. Chips are deliberately produced for the production of plywood, chipboard and wood-wool for packing.

Woodworking can be divided into two main groups: machining with or without cutting (Fig. 1.1).

The machining without cutting is confined only to specific areas. Plywood cutting and burning designs on surfaces can be done with laser beams. Lasers are not generally used for cutting because they use too much energy have low cutting efficiency. The shape of components can only be changed with bending. That is why an additional cutting operation is usually necessary.

The economic importance of wood, as a major renewable resource, will certainly increase the efforts worldwide to improve tool and machine design and to develop new technologies. For example, tool (saw) vibrations increase the kerf losses, reduce product accuracy, diminish surface quality and reduce tool life. Therefore, increasing attention is being paid to control tool and machine vibration to achieve more efficient cutting. Reducing tool vibration and control can be attained by appropriate tool design, membrane stress modification and process parameters.

Thin kerf sawing is a progressive method to reduce kerf losses. The optimum feed speed a very important process parameter. It is determined by the relationship among gullet capacity, feed speed, saw speed and cutting depth. It is important to avoid over or underfeeding to achieve thin-kerf sawing (Kirbach 1995).

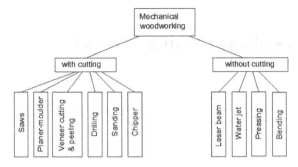

Fig. 1.1 Mechanical woodworking methods

In the following chapter we discuss the common woodworking operations and give the necessary relationships and explanations for optimum design and choice of operating parameters.

1.2 Definitions and Notations

In machining, cutting results from the interaction of the timber and the tool. The physical and mechanical properties of the timber, the geometry of the tool and the operating parameters determine the interaction. The anisotropy, the compression and the tensile strength, and the elongation at rupture are the most important material characteristics of wood. The main cutting directions corresponding to the anisotropy are demonstrated in Fig. 1.2 (Sitkei et al. 1990). When cutting with the grain, the compression strength is about the half of tensile strength, while the tensile strength perpendicular to the fibre is only about one twentieth of the tensile strength along the grain. The elongation at rupture of wood is very small compared to the metal.

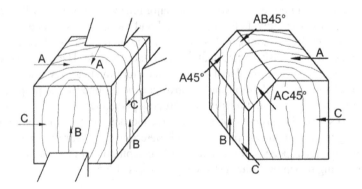

Fig. 1.2 Principal cutting directions and cutting in intermediate directions

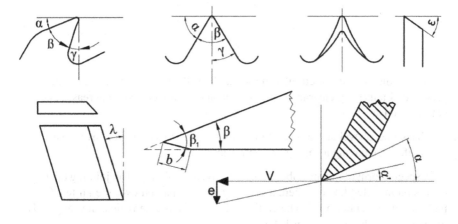

Fig. 1.3 Characteristic angles of woodworking tools

The angles and the edges are important characteristics of the tool's geometry (Fig. 1.3).

The following tool angles are important:

α The clearance angle ensures that there is no friction on the bottom face of the blade. It is generally 10°–15°, but veneer cutting knives have only around a 1° clearance angle.

β The sharpening or bevel angle; its main function is the cutting itself, maintaining a sharp edge and conducting heat away from the edge part. Veneer cutting knives have values around 20°.

γ The rake angle determines the chip deformation and is between 15° and 25°. Very hard tool materials require quite a small rake angle, while veneer cutting tools are around 70°.

ε The top bevel angle generally used on circular saws (alternately or with a double bevelled top) improve sawing accuracy.

λ The oblique or slide angle decreases the normal force on the edge and the dynamic load on long knives. An oblique angle lowers the true cutting angle according to the following equation:

$$\tan \delta' = \tan \delta \cdot \cos \lambda$$
$$\delta = \alpha + \beta$$

where $\delta = \alpha + \beta$ is the cutting angle.

α' the moving clearance angle occurs in peeling veneer and drilling due to the combine effect of a circular and a linear feed motion:

$$\tan \alpha' = \frac{e}{v}$$

where v is the velocity of circular motion and e is the speed of linear motion. In the presence of a moving clearance angle, the actual clearance angle decreases and can entirely vanish.

β_1 back microbevel angle is used in cutting veneer to avoid excessive wear of the knife edge having a small sharpening angle.

A special form of the microbevel angle is given in jointing. The jointing operation occurs on installed knives where a grinding stone is passed over the knife edges as they turn in the cutter head. The cutting edge itself has a zero clearance angle. The jointed land width is less than 1.0 mm.

The edges of the tool are important in the cutting process. We distinguish between principal and secondary cutting edges; the meeting point of the two edges is the *cutting-point*. The *cutting-edges* are never absolutely sharp, but they always have a rounding off radius. The range of the radius of the main-cutting edge is $\rho = 10\text{–}60\ \mu\text{m}$.

If the radius becomes larger due to wear, the tool edge has to be sharpened (Fig. 1.4).

The operational parameters of a tool can be summarized as follows:

- The peripheral speed of the tool (v) that influences its productiveness, the surface quality, the wear of the edges, and to a lesser extent the cutting force. Regular values vary between 30 and 60 m/s today.
- The feed per tooth (e_z), or the chip thickness (h) influence the productiveness, the surface quality and the cutting force. If the gullet of the tooth runs full in a gap, then the chip has to fit loosely in the tooth's gullet, and that limits the maximum value of e_z. We get varying chip cross sections when using rotating tools, so we should calculate the average chip thickness. Using a frame saw, the value of e_z changes as a function of the stroke, as the saw's speed is variable, while the feed motion is constant. When cutting wood, the theoretical thickness of a chip is almost the same as the thickness of the detached chip. The reason is that the

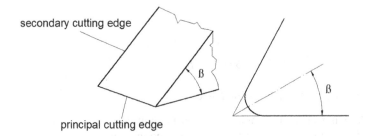

Fig. 1.4 The edges of the tool and the round edge

chip's plastic deformation is not significant; smaller plastic deformations occur only on the compressed side. When cutting metal, the thickness of the detached chip is two to three times as big as its theoretical thickness.

- The number of edges of a tool (z) influence its productivity, the surface quality, and sometimes the maximum feed per tooth. The pitch of the teeth and the capacity of the tooth gullet are important in saws.

There are relatively simple mathematical relations between the tool parameters and operational factors, which will be given in Chap. 3.

1.3 Mechanics of Cutting

The mechanics of cutting examines the interactions between the tool and the chip using the laws of engineering mechanics (Timoshenko and Goodier 1951). We take into consideration the mechanical properties of the material, if necessary the anisotropy, the viscoelasticity and its plastic properties. We look for relationships between the forces and strains (deformations), and the final goal is the determination of the cutting force depending on the properties of the material and the tool. Our aim is to provide a scientific foundation to the phenomena which so far has only been observed.

We also confirm the theoretical results with experimental data. Where necessary, we give diagrams to more quickly determine the required parameters.

1.3.1 Force and Stress Relations

Certain assumptions usually have to be made at the beginning of theoretical derivations. Our most important initial assumption concerns on the shape of the chip. We suppose that the chip deforms on the knife's surface on a radius R. A series of practical observations have confirmed this assumption. e.g. the shavings of a planer come off the tool's front surface in a closely rolled state using cutting direction B. The rolled state of the chip remains after cutting (after unloading) because the chip's compressed side has a plastic flow and this means permanent deformation.

The force and stress relations of the chip are demonstrated in Fig. 1.5 (Sitkei 1983; Sitkei et al. 1990). The normal force N acts on the tool's front surface, which can be divided into horizontal and vertical components. The chip slides on the front surface; therefore a frictional force $S = \mu N$ arises on the rake face that can be divided into components. The resultant force from the sum of the components shows an inclination towards the direction of the movement. Its angle from the vertical will be given as follows:

$$P_h = N \sin \delta + \mu N \cos \delta$$

$$P_v = N \cos \delta - \mu N \sin \delta$$

from which:

$$\tan \delta' = \frac{P_h}{P_v} = \frac{\tan \delta + \mu}{1 - \mu \cdot \tan \delta} \tag{1.1}$$

where P_h is the horizontal (cutting) force and P_v is the vertical force.

The internal stress distributions are also given in the Figure. The internal stress distribution due to bending is asymmetric because the compressive strength is roughly half of the tensile strength. The compression stress is near constant on the compressed side where the load on the material equals its strength. The result of the asymmetric stress distribution is that the neutral axis shifts to the tension side. Further significant shifting of the neutral axis to the tension side occurs because the horizontal component of the cutting force exerts an axial pressure on the chip, as on a bent beam, so it further increases the width of the compressed zone, which is loaded up to its maximum strength. (This observation is used effectively when bending furniture parts).

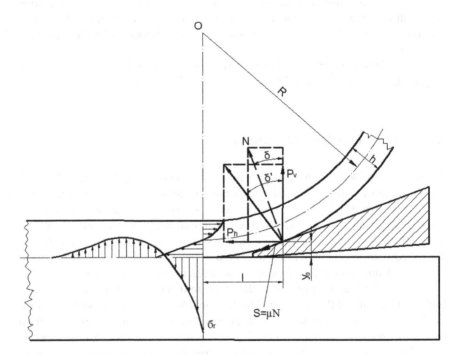

Fig. 1.5 Force and stress relations of the chip

The bending of the chip produces stresses in the horizontal level in front of the knife-edge. It is important to know this because in some cases there may be *pre-splitting of the material*. Then the surface is not actually cut by the knife edge, so we get a very rough surface similar to the surface of wood that was split by an axe.

The stress distribution in front of the cutting edge can be calculated according to the following equation:

$$\sigma_{zr} = k \cdot y$$

where:

k the deformation modulus of the wood, N/mm^3;
y the deformation in a radial direction.

The k deformation modulus can be determined approximately with an indenter. In the theory of elasticity the so-called "Boussinesq-problem" gives the following correlation (Timoshenko and Goodier 1951):

$$p = \frac{4E}{\pi\left(1 - v^2\right)} \cdot \frac{y}{d} = k \cdot y \tag{1.2}$$

where:

d the diameter of the indenter,
E the modulus of elasticity of the material in a radial direction,
v Poisson's ratio,
p pressure.

The k value depends on the diameter of the indenter according to Eq. (1.2). From our own experiments on oak, using an indenter 6 mm in diameter, we calculated 80–85 N/mm^3.

The y deformation in a distance x from the edge of the knife can be calculated as:

$$y = -\frac{P_v \cdot e^{-\beta \cdot x}}{2\beta^3 E \cdot I}[\cos \beta x + \beta \cdot l(\cos \beta x - \sin \beta x)] \tag{1.3}$$

where:

$$\beta = \sqrt[4]{\frac{k}{4EI}}$$

P_v the vertical force component,
I the moment of inertia of the chip,
l the distance of the attack point of the P_v force.

Using the above equations, the maximum tensile stress occurs under the edge of the knife, its value is given from Eqs. (1.2) and (1.3) as:

$$\sigma_{zr\,max} = \frac{kP_v(1 + \beta \cdot l)}{2\beta^3 \cdot E \cdot I} \tag{1.4}$$

We can see the change of radial deformation in a given set of parameters ($h = 0.5$ mm, $\delta = 50°$, $h/R = 0.25$) in Fig. 1.6. The deformation in a given distance x changes from tensile to compressive. Its relative location (x^*) can be seen in Fig. 1.7 depending on the thickness of the chip. (Sitkei et al. 1990).

As the thickness of the chip increases, the relative zero point approaches the edge of the knife, and this implies that the danger of pre-splitting increases as the thickness of the chip increases.

The conditions of pre-splitting can be formulated as follows:

1. The maximum value of the radial load σ_{zrmax} has to be larger than the tensile strength of the wood.
2. The chip may not break, that is that the load on the tensile side of the chip has to be smaller than the tensile strength of the wood.

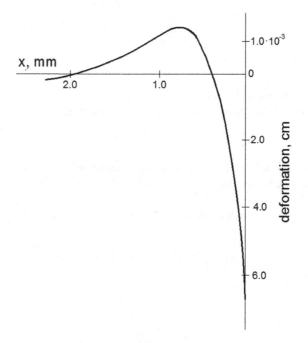

Fig. 1.6 The changing of radial deformation in the plane in front of the edge

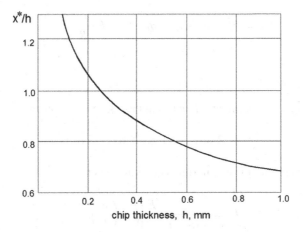

Fig. 1.7 The displacement of the relative zero point in the plane in front of the edge depending on the chip thickness

1.3.2 The Cutting Force

In order to determine the components of the cutting force, we use the balance of the moments. If the neutral axis lies in the middle of the beam, then the expressions of the relative elongation and the resistance moment are:

$$\varepsilon_t = \frac{r - R}{R} = \frac{h}{2R} \quad \text{and} \quad W = \frac{bh^2}{6}$$

These value alter when the neutral axis is dislocated in the following way: on the tension side the strain decreases

$$\varepsilon_t = \frac{h}{5 \cdot R}$$

while on the compression side it considerably increases

$$\varepsilon_c = \frac{h}{1.25R}$$

The resistance moment W also decreases in the folollowing extent

$$W = \frac{bh^2}{10}$$

Using the above approximations, the equilibrium moment equation can be given as follows:

$$M = P_v l + P_h \cdot y_0 = \sigma_t \cdot W = E \frac{h}{5R} \cdot \frac{b \cdot h^2}{10} \qquad (1.5)$$

where:

b the width of the chip.

The values of the coordinates l and y_0 in the above equation are approximately:

$$l \cong R \sin \delta \quad \text{and} \quad y_0 \cong R(1 - \cos \delta)$$

Inserting these into the equilibrium equation we get:

$$P_v = \frac{1}{f(\mu, \delta)} \cdot \frac{E \cdot b}{50} \cdot \left(\frac{h}{R}\right)^2 \cdot h \qquad (1.6)$$

where:

$$f(\mu, \delta) = \sin \delta (1 - \mu \cdot \tan \delta) + (\sin \delta - x_0/R) \cdot \tan \delta \cdot (\tan \delta + \mu)$$

and the x_0 in the last equation means the geometrical pre-splitting (see Fig. 1.10).

Considering Eq. (1.1), the resultant horizontal force, the cutting force can be calculated from the following equation:

$$P_h = \frac{\tan \delta'}{f(\mu, \delta)} \cdot \frac{E}{50} \cdot b \cdot \left(\frac{h}{R}\right)^2 \cdot h \qquad (1.7)$$

Equation (1.7) gives the force which is spent to the deformation of the chip. In reality the rounded edge is affected by force which has to overcome the compressive strength of the material. This force is proportional to the cross section of the edge and the compressive strength of wood:

$$P_h' = s \cdot b \cdot \sigma_c$$

where:

$s \cong 2\rho$ the thickness of the edge,
ρ the radius of the edge,
σ_c the compressive strength of the wood.

The final expression of the cutting force related to the unit width will be as follows:

$$P_h/b = 2\rho\sigma_c + \frac{\tan \delta'}{f(\mu, \delta)} \cdot \frac{E}{50} \cdot \left(\frac{h}{R}\right)^2 \cdot h \qquad (1.8)$$

Examining the second term in the above equation, we will come to important conclusions.

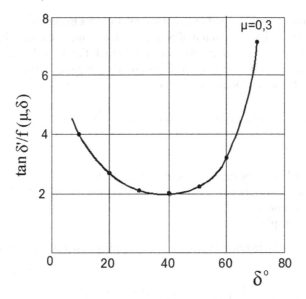

Fig. 1.8 The minimum of the cutting force depending on the δ angle (Sitkei 1983)

The $\tan \delta' / f(\mu, \delta)$ trigonometric function occurring in the second term of Eq. (1.8) is illustrated in Fig. 1.8. One can see that the diagram has a minimum at about $\delta = 40°$, and that means that the cutting force is minimal in the given range. This statement is totally in accordance with our experimental observations. Using other friction coefficients only slightly modify the diagram and the location of the minimum.

The values of friction coefficients measured at low surface pressures cannot be used here. The surface pressures are on the order of 100 bar, and here the friction coefficient is smaller. The effect of higher water content is different from the usual one. Water comes to the surface as a result of the pressure, and this lowers the friction coefficient. The high temperatures on the friction surface also considerably decrease the friction coefficient. Therefore we record the values of μ between 0.25 and 0.35.

The Eq. (1.8) is linear depending on the thickness of the chip, so far as the h/R ratio is constant at a given δ angle. According to experimental results, Eq. (1.8) is linear with a good approximation in any cutting direction; therefore, it is worthwhile to further study the deformation of the chip.

1.3.3   Chip Deformation

In the previous derivations, we supposed that the point of attack of the cutting force and therefore the h/R ratio is known. That is not the case. Chip deformation has more than one degree of freedom and therefore, the h/R ratio cannot be calculated from static equilibrium equations. The chip can deform in different ways depending on the combined action of internal and external forces. In order to determine the h/R ratio, the variational principle (minimum of potential energy) may be used. The basic idea of this method is to find a chip deformation in which the work done by the forces would be a minimum. To solve this problem, we describe the virtual work produced by internal and external forces as follows

$$\delta(U_i + U_e) = 0 \quad \text{and} \quad U_e = -\sum F_j \delta y_j$$

where

U_i, U_e the work done by internal and external forces,
δ symbol of variation,
F_j external forces,
δy_j virtual displacements.

The work done by internal forces in a bent beam is given by the following equation (Ritz method):

$$U_i = \frac{1}{2} \int E \cdot I y''^2 dz = \frac{1}{4} \left(\frac{\pi}{2l}\right)^4 \cdot E \cdot I \cdot y^2 \cdot l$$

where y'' means the curvature of the neutral axis.

The work done by external forces is expressed in the following form:

$$U_e = \frac{P_h \Delta l}{2} + \frac{P_v \cdot y_0}{2}$$

in which the longitudinal deformation of the chip is given by

$$\Delta l = \frac{P_h \cdot l}{E \cdot b \cdot h}$$

Keeping in mind Eq. (1.1), the tangential and radial force components can be expressed using the bent beam theory:

$$P_h = \frac{3EI y_0}{l^3}(\tan \delta + \mu)$$

$$P_v = \frac{3EI y_0}{l^3}(1 - \mu \tan \delta)$$

Taking the functional $F = U_i - U_e$ and its derivative with respect to y, setting the functional equal to zero, we get

$$\frac{\partial F}{\partial y} = 0$$

For the length of the beam we may take the expression

$$l \cong R \cdot \sin \delta$$

or the more accurate

$$l = R \cdot \delta$$

relationships. In the latter case the cutting angle δ must be substituted in radian.

Solving the above equation and using the $l \cong R \sin \delta$ approximation, the h/R ratio is given by the following equation:

$$\frac{h}{R} = \frac{2 \sin \delta}{(\tan \delta + \mu)} \sqrt{\mu \tan \delta} \qquad (1.9)$$

In the above derivation we have supposed that the total bending work occurs in the elastic range and no plastic deformation occurs, but that is not the case. Beyond the yield limit, the bending forces remain constant and do not increase linearly with the deformation. Closer examination of this distortion gave the following correction:

$$\frac{h}{R} = \frac{2 \sin \delta}{\sqrt{3}(\tan \delta + \mu)} \sqrt{\mu \tan \delta} \qquad (1.9a)$$

We can see from the equation that the h/R ratio depends only on δ and μ, and these are constant for a given tool. It follows that Eq. (1.8) is linear depending on the h chip thickness. In simpler terms:

$$P_h/b = A + B \cdot h \qquad (1.8a)$$

Figure 1.9 graphically demonstrates Eq. (1.9a), where we have marked measurement results (Sitkei 1983). The values of the measuring results are systematically smaller than those obtained theoretically. This can be explaining by the fact that the upper side of the chip yields plastically, and this leads to the rotation of the bent cross section, thus reducing the apparent value of the δ angle (Fig. 1.10). Therefore the examination of the chip deformation has to be expanded in the plastic state.

1.3.4 Chip Deformation with Plastic Flow

Due to the high load, the upper, compressed side of the chip practically always yields. (The load reaches and surpasses the strength of the chip). As a consequence, the cross section at the foot of the chip turns in a forward direction, and the δ angle apparently decreases (δ^*). Figure 1.10 represents the rotation of the cross section and the decreasing of the geometric pre-splitting (x_0). *Geometric pre-splitting* begins

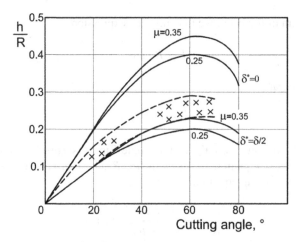

Fig. 1.9 *h/R* ratio as a function of cutting angle for two friction coefficients (dotted line is the boundary for measurement data)

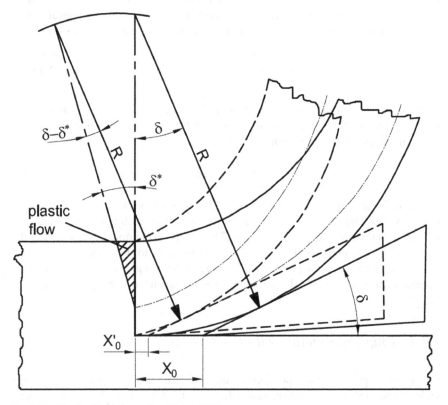

Fig. 1.10 Plastic chip deformation (Sitkei 1983)

because the plane surface of the tool and the arc-shaped chip touch at one point. We can see from the Figure that the value of x_0 decreases quickly as the δ^* angle increases. This decreased value of x_0 can decreases further, and can cease completely so that the convex surface of the chip locally deforms, and the theoretical contact line turns into a contact surface.

The theoretical h/R ratio can also be calculated taken the effect of plastic chip deformation into account. In this case Eq. (1.9a) is modified to the following form:

$$\frac{h}{R} = \frac{2 \cdot (\delta - \delta^*)\sqrt{\mu \cdot \tan \delta}}{\sqrt{3}(\tan \delta + \mu)} \tag{1.9b}$$

The results of calculations are also given in Fig. 1.9 with the assumption that the rotation angle δ^* is the half of the cutting angle. In this case the theoretical value agrees well with the measured values.

As we see in Fig. 1.10, without plastic deformation the geometric pre-splitting x_0 has a considerable magnitude which would result in a split uneven surface. Its length depends on the radius of deformation and the cutting angle and is calculated by the following equation

$$x_0 = R \cdot \left(\sin \delta - \frac{1 - \cos \delta}{\tan \delta}\right)$$

The plastic deformation of chip with rotation angle δ^* considerably decreases the geometric pre-splitting and its value as a function of rotation angle is calculated as follows

$$\frac{x_0'}{R} = \frac{x_0}{R} - \frac{\sin \delta^*}{\sin \delta}\left[\frac{1}{\sin \delta} - \cos \delta\left(\tan(90 - \delta) + \frac{1 - \cos \delta^*}{\sin \delta^*}\right)\right]$$

The equilibrium value of rotation angle δ^* is an important question. Because the cutting occurs with high velocity, therefore, dynamic effects dominate. This dynamic equilibrium can hardly be calculated reliably and no experimental observations exist. We suppose that a static equilibrium may be in the range of around $\delta^* = 0.5\delta$. Local flattening of the chip convex surface further decreases the pre-splitting and can even cease it due to dynamic forces transmitted by the tool.

The coordinates of the resultant force with plastic chip deformation are given in the following form

$$l = R(\sin \delta - \sin \delta^*)$$
$$y_0 = R(\cos \delta^* - \cos \delta)$$

In following we have to determine the geometric conditions of pre-splitting. In Fig. 1.11, the chip tensile stress due to bending σ_{lh} and the radial tensile stress in the

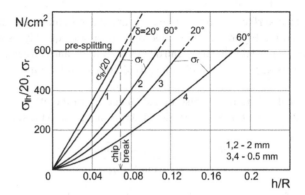

Fig. 1.11 Chip tensile stress due to bending (σ_{lh}) and radial tensile stress in the separation zone (σ_r) as a function of the h/R ratio for two different chip thicknesses. (To the explanation of the pre-splitting phenomenon before the knife edge) (Sitkei 1983)

separation zone σ_r are demonstrated as a function of the h/R ratio for two different chip thicknesses.

Because tensile strength of wood in the direction of the grain is approximately 20 times larger than the radial tensile strength, we have reduced it by 20 ($\frac{\sigma_{lh}}{20}$). The radial tensile stress depends on the cutting angle and chip thickness, so we have constructed several σ_r curves to demonstrate the effect of these variables. Pre-splitting will obviously happen, if the σ_r curve for a given case reaches the horizontal strength boundary line before the chip would break. As we can see from the Figure, this will generally be realized only for chips that are $h \geq 2$ mm thick. The use of smaller cutting angles is more inclined to cause pre-splitting. For making veneers, we use a cutting angle around 20°. Therefore, the use of a pressure bar is always essential to hamper pre-splitting.

In the practice we rarely produce chips thicker than 2 mm. Therefore, pre-splitting seldom happens.

Under steady state cutting condition the coordinates of the resultant force does not change and the cutting force is constant. In practice, due to wood inhomogeneities, the cutting force is always changing around an average value. Change in chip thickness causes cutting force variations as it is typical for rotating cutting tools.

Making high speed wood machining, the chip is generally comminuted into small fragments with a given size distribution. Reasons for this are the large strains during chip formation and the dynamic load on the chips. The maximum strains can easily be predicted using the simple strain equations. With $h/R = 0.25$, we get

$$\varepsilon_c = \frac{h}{1.25 \cdot R} = 0.2 \quad \text{and} \quad \varepsilon_t = \frac{h}{5 \cdot R} = 0.05$$

which are well over the yield limit both for compression and tension. An entire ribbon-like chip is produced by using hand planer with sharp edge. The cleanest cut surface is made by Japanese planer (see in Chap. 8).

1.3.5 The Radial Cutting Force

In the derivation of Eq. (1.8) we have supposed for simplicity that the resultant force on the edge is horizontal, (parallel with the direction of cutting). In reality the stress distribution on the edge land is not symmetric, so the resultant force closes up an angle with the direction of motion. The reason is that the extension of the material under the edge of the tool is relatively big, while only an h thickness material is above the edge, (the open wall problem).

The direction of the resultant force is not known exactly and it also depends on the thickness of the chip. Therefore we take, as a first approximation that the direction of the resultant force falls into the bisector of the edge of the knife (Fig. 1.12).

The resultant force, acting on the edge is [see in Eq. (1.8)]:

$$P_e \cong 2\rho b \sigma_c$$

and its components can be written as:

$$P_{eh} = P_e \cos(\beta/2 + \alpha) \quad \text{and} \quad P_{ev} = P_e \sin(\beta/2 + \alpha)$$

We can write the vertical component of the whole resultant force acting on the tool as follows:

$$P_v/b = 2\rho\sigma_c \sin(\beta/2 + \alpha) - \frac{B \cdot h}{\tan \delta'} \tag{1.10}$$

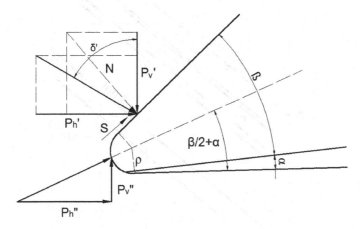

Fig. 1.12 The position of the force acting on the edge (Sitkei 1983)

The horizontal component has the form:

$$P_h/b = 2\rho\sigma_c[\cos(\beta/2 + \alpha) + \mu \cdot \sin(\beta/2 + \alpha)] + B \cdot h \qquad (1.11)$$

where:

$$B = \frac{\tan \delta'}{f(\mu, \delta)} \cdot \frac{E}{50} \cdot \left(\frac{h}{R}\right)^2$$

The numerical evaluation of Eq. (1.10) is shown in Fig. 1.13 (Sitkei et al. 1990). The vertical component changes its sign depending on the thickness of the chip and the radius of the edge. The force component is practically always negative (tension) when chips are large. The phenomen has been observed in the early 1960s by measurements with a pendulum dynamometer (Sugihara et al. 1966).

The ratio of the two components can be determined from the derived equations. We can see the P_v/P_h ratio as a function of edge radius in Fig. 1.14. (In both Figures $\beta = 60°$, $\alpha = 15°$ and $\sigma_c = 600$ daN/cm^2.) We also see the change of the sign.

It is known from measurements that by using rotating tools, the radial force component as a function of angular motion has a different shape depending if the tool is sharp or blunt (Fig. 1.15). This phenomenon can now be explained clearly in

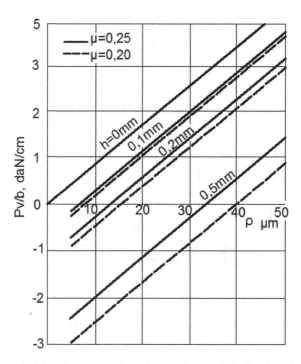

Fig. 1.13 The vertical force component depending on the rounded off radius of the edge (Sitkei et al. 1990)

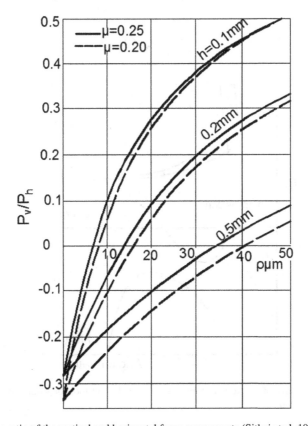

Fig. 1.14 The ratio of the vertical and horizontal force components (Sitkei et al. 1990)

Fig. 1.13. The thickness of the chip always alters when using rotating tools. If the tool is sharp, then chips 0.2–0.3 mm thick produce a negative radial force and the force turns into a positive value as the chip thickness decreases. Blunt tools almost always produce a positive radial force.

The radial force is always counteracted by the guiding and bearing of the tool. Large positive radial forces (from blunt tools) load the bearing and decreasing its lifetime, but they also have other effects. The positive radial force is an exciting force that induces the vibration of the work-piece and the rotating tool. The vibrations worsen the quality of the surface (see Sect. 5.5).

In practice, radial forces increase as a consequence of edge wear and jointing. Jointing is a common practice to produce the same cutting circle for all knives of a cutterhead or a peripheral knife planer. Jointing allows several re-sharpening without removing the knives from the cutterhead. The jointed land at the cutting edge has a zero degree clearance angle which can compress a thin layer of the planed board causing cell damage and heat generation due to excessive friction. Measurements show that the jointing increases the radial force component as jointing land width increases. (Jointed land width is generally less than 1.0 mm.)

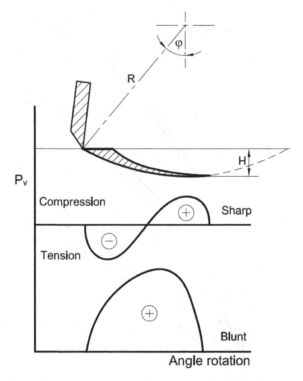

Fig. 1.15 The course of the vertical force for sharp and blunt tools depending on the angle of rotation (Sitkei 1983)

1.4   Mechanics of Oblique Cutting

The oblique cutting of materials in the metal and wood industry and also in agriculture has been widely known and used for long time (Sitkei 1986).

The main advantages of oblique cutting can be summarized as follows:

- the cutting force can considerably be reduced. If the possible counterbalance force of the material is low compared to the cutting force, then the slide cutting method can only be used on soft plastic and upholstering materials.
- the reduction of normal force always means a smaller dynamic load in the driving and bearing mechanisms and also a smaller excitation force generating workpiece vibration.
- one of the main advantage of oblique cutting is lowering the noise level, which may be as high as 6–8 dB.
- it may also be assumed that oblique cutting produces a better surface quality.

It seems to be advisable to express the force reduction function in the following dimensionless form (Sitkei 1986):

$$(P_0 - P_\lambda)/P_0 = f\{(\delta - \delta')/\delta\} \qquad (1.12)$$

where P_λ is the force at a given oblique angle λ and δ' are the effective cutting angle at a given λ. The cutting energy also varies as a function of oblique angle and it shows a minimum of about $\lambda = 20°–25°$ (Sitkei 1997).

The kinematic relation of the oblique cutting is given in Fig. 1.16. The cutting edge is not orthogonal in the direction of motion but it has an oblique angle (slide angle) λ and the resultant speed consists of two components:

$$\tan \lambda = \frac{v_t}{v_n} \quad \text{and} \quad v_t = v \cdot \sin \lambda, \quad v_n = v \cdot \cos \lambda, \qquad (1.13)$$

The value of $\tan \lambda$ is often termed the sliding coefficient. The effective cutting angle δ' varies as a function of the oblique angle according to the following equation:

$$\tan \delta' = \tan \delta \cdot \cos \lambda \qquad (1.14)$$

With increasing slide angle, the effective cutting angle can considerably be reduced.

The cutting force consists of two components: one acts on the round edge surface and the second one is due to chip deformation [see Eqs. (1.8) and (1.8a)]. The ratio of these components varies with chip thickness (Fig. 1.17).

The force component acting on the edge depends on the edge radius (dullness) and on the compressive strength of the material. The force due to chip deformation depends on the cutting angle and chip thickness. One of the most characteristic features of oblique cutting is the presence of shear stresses in the separation zone. It is well known that shear stresses contribute to the failure of materials and the equivalent stress is given by the relation:

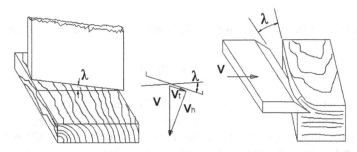

Fig. 1.16 Kinematic relations of oblique cutting and penetration of the knife into the material (Sitkei 1987)

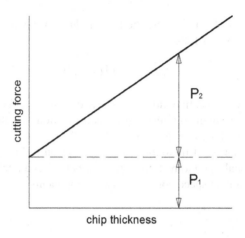

Fig. 1.17 Cutting force versus chip thickness

$$\sigma_e^2 = \sigma^2 + (2 \cdot \tau)^2 \tag{1.15}$$

In cutting, the shear stress is proportional to the normal stress and to the friction coefficient, that is $\tau = \mu\sigma$. The work done by the shear deformation is proportional to the tangential speed component Therefore, the friction power on a given surface F is expressed by the relation:

$$W_{fr} = \mu\sigma \cdot F \cdot v \cdot \sin \lambda \tag{1.16}$$

Keeping in mind Eq. (1.15), the force acting on the edge can be expressed as a function of oblique angle in the following manner:

$$\sigma_\lambda \cdot F = \sigma \cdot F - 2\mu\sigma \cdot F \cdot \sin \lambda$$

or

$$P_\lambda = P_0(1 - 2\mu \cdot \sin \lambda) \tag{1.17}$$

Using Eq. (1.17), the relative force reduction for the first force component will be

$$\frac{(P_0 - P_\lambda)}{P_0} = 2\mu \cdot \sin \lambda \tag{1.18}$$

The second force component (P_2 in Fig. 1.17) is due to chip deformation. The following relations are known from the theory of wood cutting (Sitkei 1983): the force is proportional to the square of the relative chip deformation h/R and h/R is proportional to the 0.75th power of the cutting angle, that is:

$$P \sim (h/R)^2 \quad \text{and} \quad h/R \sim \delta^{0.75} \quad \text{therefore } P \sim \delta^{1.5}$$

Keeping in mind the above discussion, the relative force reduction of the second component is:

$$\frac{(P_0 - P_\lambda)}{P_0} = \frac{(\delta^{1.5} - \delta'^{1.5})}{\delta^{1.5}} \tag{1.19}$$

Equations (1.18) and (1.19) are plotted in Fig. 1.18. If we assume for simplicity that both components have the same magnitude, then the resultant force reduction will be given by the geometric mean calculated from the components. This line (R) is also plotted in Fig. 1.18. These considerations show that force reduction is not uniquely determined by reducing the cutting angle. To a certain extent, the force reduction will also be influenced by the cutting angle (P_1 component) and by the P_1/P_2 ratio (resultant force).

The penetration of the edge into the material varies considerably if oblique cutting is used (Fig. 1.19). As a consequence of the inclined position, the chip will be lifted from one side before cutting takes place and, therefore, the cutting occurs under the combined action of tensile, compressive and shear stresses. The stretching tensile stress acting vertically to the shear plane reduces the compressive stress under the edge during cutting and produces a better surface quality. Oblique cutting transforms the rounded edge into the section of an ellipse having a smaller curvature radius than that of the edge ($\rho' = \rho \cos \lambda$).

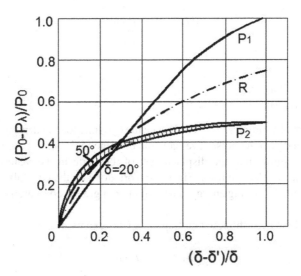

Fig. 1.18 Theoretical force reduction as a function of reducing the cutting angle (Sitkei 1997)

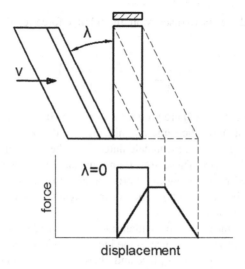

Fig. 1.19 Force displacement relation for different slide angles (Sitkei 1997)

A further advantage of the oblique cutting is a smaller rate of increase of force at the start of cutting (Fig. 1.19). The amount of force increase for rotating tools is given by the following equation:

$$\frac{dP}{dt} = \frac{P_\lambda \cdot R \cdot \omega}{\tan \lambda} \qquad (1.20)$$

where

R is the tool radius,
ω is the angular speed.

The amount of lifting force will influence the dynamic load in driving and bearing mechanisms and also the noise level generated by the tool.

Figure 1.19 shows that the displacement of the knife in the material varies as different oblique angles are used. Since the work done by the edge also depends on its displacement, the cutting energy can be smaller or greater compared to orthogonal cutting.

The work done by orthogonal cutting ($\lambda = 0$) is expressed as:

$$A_0 = P_0 \cdot L \cdot b$$

In oblique cutting

$$A_\lambda = P_\lambda \frac{b^2}{\sin \lambda} + P_\lambda \frac{b}{\sin \lambda}(L \cdot \tan \lambda - b) = \frac{P_\lambda \cdot L \cdot b}{\cos \lambda}$$

The ratio of the work performed by these two cutting methods is expressed by

$$\frac{A_\lambda}{A_0} = \frac{P_\lambda}{(P_0 \cos \lambda)} \tag{1.21}$$

where

L is the length of cutting,
b is the width of cutting.

Measurement series for a given oblique angle and cutting angle will be demonstrated as a function of chip thickness. Figure 1.20 shows the results in veneer cutting range. The specific cutting force is linearly related to the chip thickness as predicted theoretically. The curves for different oblique angles show a definite influence of the sliding motion on cutting forces.

The specific force reduction was calculated from the measured cutting forces and the functional relationship was established according to Eq. (1.12). This experimental relationship is plotted in Fig. 1.21. The curve obtained experimentally is surprisingly similar to Fig. 1.18, which is worked out theoretically. The slight scattering of experimental results is explained by the reasons mentioned earlier in Fig. 1.18.

The general relationship given in Fig. 1.21 is expressed by the following half-empirical equation:

$$\frac{(P_0 - P_\lambda)}{P_0} = 0.75 \left\{ \frac{(\delta - \delta')}{\delta} \right\}^{0.6}$$

The relative cutting energy can be calculated by using the measured cutting forces and Eq. (1.21). These results are shown in Fig. 1.22. The optimum oblique angle from an energetic point of view is a function of the cutting angle and for common cutting

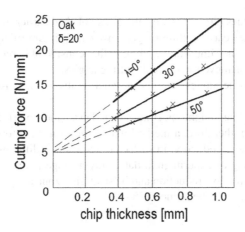

Fig. 1.20 Cutting force versus chip thickness for different oblique angles (Sitkei 1997)

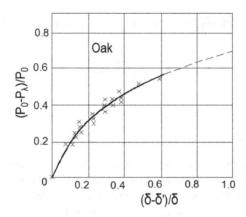

Fig. 1.21 Dimensionless relationship describing force reduction (Sitkei 1997)

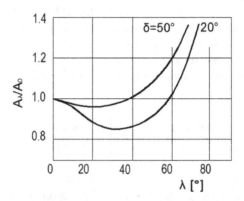

Fig. 1.22 Cutting energy versus oblique angle at two cutting angles (Sitkei 1997)

angles the same values (20°–25°) are valid as established earlier for farm machinery (Sitkei 1986). In the veneer cutting range, however, optimum values are obtained at about $\lambda = 35°$. Another important finding is that considerable reduction in cutting energy can only be achieved with small cutting angles.

Using large oblique angles, the cutting energy always increases due to excess friction.

The surface roughness of the specimens varied in a wide range (R_z values between 40 and 90 μm) and, therefore, a method which shows more information was used to calculate the relative reduction in surface roughness for different oblique angles. Because of scattering of experimental data, only average reduction values may be given. The average relative reduction and its scattering zone are plotted in Fig. 1.23. In practice, roughness can be reduced 15–20%.

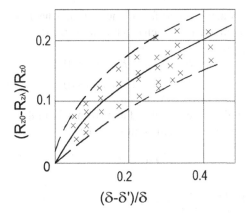

Fig. 1.23 Relative surface roughness reduction as a function of reducing the relative cutting angle (Sitkei 1997)

1.5 Mechanics of Bundle Clipper

Clipping a veneer bundle is a special case in woodworking. The bundle is generally 4–6 cm high. The bundle is pressed down with a nose bar before the start of the cutting. Figure 1.24 shows the force relations in the cutting material. The force is composed of two main parts:

- the force acting on the edge and
- the force from deformation and shifting of the material.

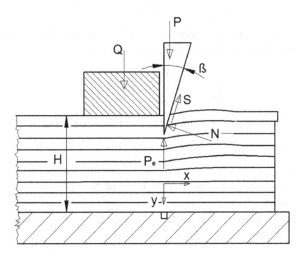

Fig. 1.24 The force-relations of the bundle clipper

The normal force N acting on the face of the knife and the frictional force $S = \mu N$ can be divided into components which can be summarized:

$$P_h = N(\cos \beta - \mu \sin \beta)$$
$$P_v = N(\sin \beta + \mu \cos \beta)$$

The normal force N can be calculated approximately in the following way. The elemental vertical force related to the unit width is:

$$dP_v = E\varepsilon dx$$

and keeping in mind that

$$dx = \tan \beta dy \quad \text{and} \quad \varepsilon = \frac{h}{H}$$

where h is the deformation within one layer, we get

$$P_v = \left(\frac{E \tan \beta}{H} \right) \int_0^{h_i} h dy = \frac{E h_i^2}{2H} \tan \beta \qquad (1.22)$$

Knowing the vertical force component due to deformation, the normal force N on the face of the knife will be

$$N = \frac{E h_i^2}{2H \cos \beta}$$

Due to the sliding motion, the normal force N generates a friction force and the total vertical force due to deformation has the final form

$$\sum P_v = \frac{E h_i^2}{2H} (\tan \beta + \mu)$$

The force component acting on the edge can be calculated as

$$P_e = 2\rho b \sigma_c$$

where ρ is the edge radius, b is the width (length) of the knife and σ_c is compressive strength of the wood.

The total vertical force on the knife is the sum of these two force components which yields

$$P_e = 2\rho b \sigma_c + \frac{E h_i^2}{2H} (\tan \beta + \mu) \qquad (1.23)$$

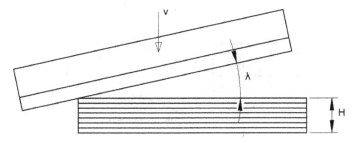

Fig. 1.25 Bundle clipper knife with its slide angle

When the upper layer is cut, this layer will be shifted aside and it exerts no resistance to the knife.

The bundle clipper also has a sliding angle λ (oblique cutting, Fig. 1.25) and the length of the cut is $L = H / \sin \lambda$.

1.6 Mechanics of the Pressure Bar

Veneer is generally produced by plane cutting or peeling. In order to produce high quality veneer, the peeling machine, especially its knife and nose bar, should always be set in its optimum position. Furthermore, the wood material will be cut when wet and heated ensuring more plasticity to avoid surface checking (cracking).

The main factor governing surface checking is the radius of deformation of the veneer on the rake face of the knife. Therefore, the cutting angle of the knife ($\delta = \alpha + \beta$) is as small as possible, generally 20°. The clearance angle varies between 0.75° and 1.5°. The small bevel angle makes the edge very sensitive to wood defects. In some cases, an attempt was made to decrease the tool edge sensitivity by using a microbevel angle of 25°–30° (see in Fig. 1.3 and later in Sect. 3.8).

An important measure in quality veneer production is the use of a pressure bar (nose bar) setting over the knife in an appropriate position (Fig. 1.26). The deformation and the position of the nose bar are related to the veneer thickness in the following way:

The relative deformation is z_0/h; the horizontal distance between the rear edge of the nose bar and the cutting edge related to the veneer thickness c/h is the *relative lead* (see Fig. 1.26).

The pressure exerts a vertical stress field in the wood lying under the bar. This vertical stress has beneficial effects in different ways: it diminishes the splitting of the wood in the plane in front of the cutting edge ensuring a pure cut surface by the tool edge. Furthermore, it may also be assumed that under compression stresses the cutting process is more stable producing smoother surface.

The pressure bar contacting the wood surface may have many different shapes. Some possible shapes are given in Fig. 1.27. Because the pressure bar slides in one

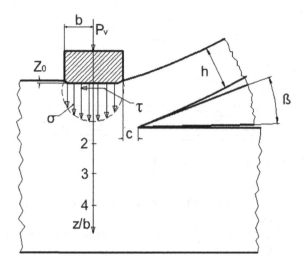

Fig. 1.26 The mechanics of a pressure bar (Sitkei et al. 1990)

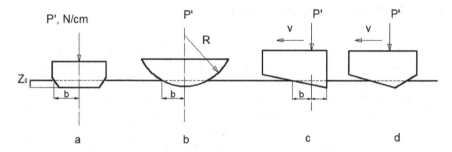

Fig. 1.27 Possible contact shapes for pressure bars

direction, it should have rounded edges or a triangle cross section. In order to exert a given vertical stress, the pressure bar should have either a prescribed deformation on the wood surface or a given load per unit length (N/cm).

The prescribed deformation is generally related to the veneer thickness and it is 10–15% of the veneer thickness. The load per unit length is 200–400 N/cm.

The pressure distribution under the nose bar can be calculated as the "Boussinesq problem", i.e. the pressure distribution in the elastic infinite half-space (Timoshenko and Goodier 1951). The contact area can be assumed as infinitely long with a half width b. That is the case of a strip load with different pressure distributions on the contact surface, as shown in Fig. 1.28. The Figure shows three different pressure distributions on the contact area and the pressure distribution on the centre line under the contact area as a function of relative depth where stress is exerted (z/b).

For a uniform strip load, the vertical stress component at a given point C is given by the equation (see Fig. 1.28):

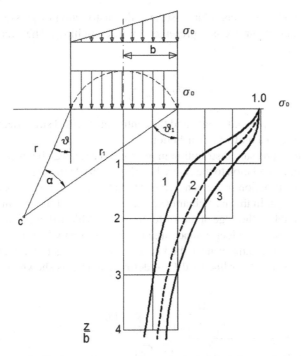

Fig. 1.28 Pressure distribution on the centre line under an infinitely long strip load. 1—triangular pressure distribution; 2—parabolic distribution; 3—constant pressure (Timoshenko and Goodier 1951)

$$\sigma_z = \frac{\sigma_0}{\pi}[(\vartheta - \vartheta_1) + \sin\alpha] = \frac{\sigma_0}{\pi}(\alpha + \sin\alpha)$$

On the centre line, the angle α can be calculated as

$$\frac{\alpha}{2} = arctg\left(\frac{b}{z}\right)$$

and therefore

$$\sigma_z = \frac{\sigma_0}{\pi}\left[2arctg\left(\frac{b}{z}\right) + \sin\alpha\right] \tag{1.24}$$

which is represented graphically in Fig. 1.28. It is important to note that the vertical stress quickly decreases as a function of depth and the relative depth z/b must always be used.

When a cylinder contacts a plane surface the distribution of pressure along the width of the contact surface is represented by a semi-ellipse. The average contact pressure is:

$$\sigma_0 = \frac{2P'}{\pi \cdot b}$$

where P' is the load per unit length and the centre line vertical stress decreases more rapidly compared to a uniform distribution.

With a triangle pressure distribution in the contact area, we obtain an asymmetric stress distribution in relation to the centre line (Fig. 1.29). In the Figure we can see the vertical stress distribution in a given ($z/b = 1$) relative depth and a possible position of the peeling knife. In this particular case only some 12% of the maximum contact pressure is available at the edge of the knife. If we would like to have a greater vertical pressure on the knife, a wider contact area must be chosen which means a smaller relative depth for the same veneer thickness. A wider contact area will simply be achieved by using a greater line load P' or deformation z_0, as shown in Fig. 1.27.

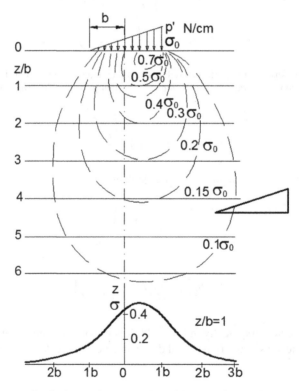

Fig. 1.29 Pressure distribution under a nose bar with triangular loading shape and pressure distribution in a plane located at $z/b = 1$

If we consider a pressure bar profile shown in Fig. 1.27c with an inclination angle of 15°, the following approximations are valid:

$$P' = 4b\sigma_0 \quad \text{or} \quad P' \cong 0.4665 \frac{Ez_0}{\left(1 - \upsilon^2\right)}$$

and

$$\sigma_0 \cong 0.1166 \frac{E}{1 - \upsilon^2} \cdot \frac{z_0}{b}$$

where E is the modulus of elasticity in the direction of normal pressure and υ is Poisson's ratio. Selection of the appropriate modulus of elasticity may give some trouble. The pressure bar slides on a cut surface with roughness and the small deformation z_0 comparable to the surface roughness. Therefore, the actual value of E may be much less than the measured value on a specimen.

The pressure bar slides on the wood surface and, depending on the friction coefficient, shear stresses develop in the contact area. This generates a force component in the direction of motion $P_h' = \mu P'$. The total peeling force will be the sum of the cutting force and the friction force.

A log is peeled in tangential direction perpendicular to the grain. Wood has little tensile strength in this direction and therefore, small tensile stresses in this plane can cause checks or cracks. At the front edge of the nose bar the wood surface suffers compression stresses, but at the rear of the contact tensile stresses occur (Fig. 1.30) (Hamilton and Goodman 1966). This tensile stress depends on the friction coefficient and on the maximum normal stress σ_0. Besides, it is wise to make the rear edge rounded to decrease the peak of the tensile stress.

Thin veneer will be produced by using oblique cutting. As explained in Sect. 1.4, cutting with a slide angle λ reduces the normal stresses on the cutting edge and the shear stresses contribute to the cutting process. In this case the total peeling force can be approximated as follows:

$$P_h = A + B_0 \cos \lambda \cdot h + \mu 2b\sigma_{av} \tag{1.25}$$

where A and B_0 are the constants in the cutting force equation for $\lambda = 0$ and σ_{av} is the average pressure on the nose bar. The μfriction coefficient has values between 0.32 and 0.38.

Similar test results for oak wood are seen in Fig. 1.31 using an oblique angle $\lambda = 50°$ and a flat nose bar with rounded edges having the half width $b = 0.5$ mm.

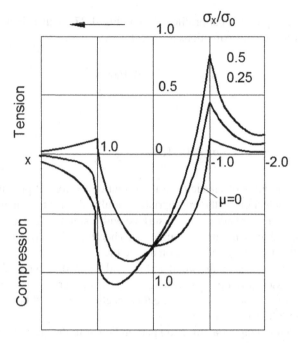

Fig. 1.30 Distribution of shear stresses under a flat nose bar for different friction coefficients (Hamilton and Goodman 1966)

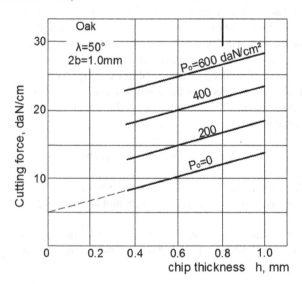

Fig. 1.31 Cutting force using a nose bar with different normal pressures (Sitkei et al.)

1.7 Comminution of Wood with Hammer Mill

The hammer mill, as a simple grinding device, was developed almost 100 years ago for fodder preparation from agricultural crop products (barley, maize) and forage (alfalfa). Great advantage of a hammer mill is its simple construction and general use for grinding of quite different granular materials. The wood industry has also discovered its applicability for grinding pre-chopped wood. Due to its working principle, comminution by impact load, the produced particle size distribution is not as uniform as required for wheat or coffee milling, especially over hammer velocities of 50 m/s.

The process taking place in a hammer mill are the result of complex loads. The material charged into a mill first impacts against the hammer, whereby it is broken into fairly large pieces and accelerated to velocity similar to the peripheral value. The accelerated particles impact against the surface of a screen, whereby they are further comminuted. Particles rebounding from the screen impact anew against the hammer. A contribution to comminution is also made by rubbing in the clearance between the hammer and the screen, i.e. by friction and impact among the particles.

A general condition in comminution is that the impact load must exceed the dynamic breaking strength depending on the mechanical properties of the material and is numerous cases also on the orientation of the particles during impact. Agricultural and wood materials are all viscoelastic, and their degree of elasticity varies considerably depending on the moisture content. This explains why the comminution of agricultural materials cannot by calculated theoretically, and so it is necessary to rely mainly on experimental results. Nevertheless, analysis of experimental results permits fundamental regularities to be revealed in the comminution process. It may therefore be stated that the general regularities of the comminution process by hammer mills are well established (Sitkei 1986).

1.7.1 Mechanism of Comminution

The aim of comminution is to reduce the size and increase the specific surface area of particles. Material breaks when the local load transferred by impact exceeds the breaking strength of the material and the energy transmitted is sufficient to overcome cohesive forces at the new surface created. The maximum pressure on contact surface for impact of a sphere at a velocity v on a plane surface is (Timoshenko and Goodier 1951):

$$p_{max} = 0.833 \left(\frac{E}{1 - \gamma^2} \right)^{0.8} \left(\gamma \cdot v^2 \right)^{0.2}$$

For the material to break $p_{max} \geq \sigma_B$. Substitution of this relation yields the critical impact velocity for breaking as

$$v_{cr} \cong 1.62 \frac{\sigma_B^{2.5}}{\sqrt{\gamma}} \left(\frac{1 - \nu^2}{E} \right)^2 \cong const \sqrt{\frac{\sigma_B}{\gamma}}$$

where γ means the density of particle. The above relationship shows the qualitative effect of the most important material and geometrical characteristics on the critical velocity. However, in examining the effect of the breaking stress the fact must also be take into account that the modulus of elasticity E is not independent of σ_B. If for the sake of approximation it is assumed that σ_B is proportional to E, then the critical velocity increases with the square root of σ_B. In the case of irregular bodies the local curvature may also have an important effect, and so the critical velocity also depends on the orientation of the body. The breaking stress of certain products such as wood also depends on the orientation: the material can support a higher load in the direction of the longitudinal axis than normal to it.

Figure 1.32 shows the shape of the hammer and its working principle. Due to several edges on the hammer tip, much of the particles impact on the edges of small curvature enhancing the local load considerably and lowering the critical impact velocity.

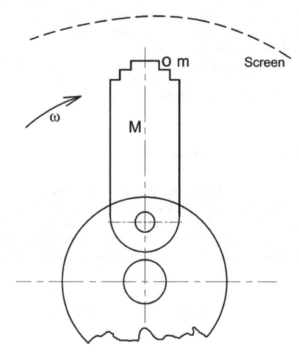

Fig. 1.32 Working principle of hammer mill

Due to the stochastic nature of impact process in the hammer mill, it appeared to be more succesful to examine the relationship between energy consumption and created new surface during comminution.

The energy required to reduce the size of a body with a given dimension x may be expressed in the general differential form:

$$\frac{\Delta E}{dx} = -\frac{k}{x^n} \tag{1.26}$$

where the constant k accounts for all the material characteristics. The exponent n generally varies between 1 and 2. If $n = 1$, then the equation due to Kick is obtained:

$$E = k_k \cdot \ln \frac{x_0}{x_1} \tag{1.27}$$

In the case of $n = 1.5$, Bond's law is valid:

$$E = k_B \left(\frac{1}{\sqrt{x_1}} - \frac{1}{\sqrt{x_0}} \right) \tag{1.28}$$

while if $n = 2$, the Rittinger's law prevails (Rittinger 1867):

$$E = k_R \left(\frac{1}{x_1} - \frac{1}{x_0} \right) \tag{1.29}$$

where x_0 is the initial grain size and x_1 that after comminution. Numerous experimental results (Bölöni 1962, 1964a, b) have shown that comminution of agricultural materials takes place according to the Rittinger equation. From Eq. (1.29) the energy required to comminute the unit volume (1 cm^3) of a material may be determined. Assume that an amount of the work C (Ncm/cm^2) is required to cut a cube of volume 1 cm^3. Smaller cubes are then obtained by three cuts. The required specific energy is

$$A = 3kC \left(\frac{1}{x_1} - \frac{1}{x_0} \right) \tag{1.30}$$

The surface area of the original cube is $6x_0^2$ and its weight is γx_0^3. The specific surface area (cm^2/kg) with these values is:

$$f_0 = \frac{6}{\gamma \cdot x_0}$$

Similarly, the specific surface area after comminution is

$$f_1 = \frac{6}{\gamma \cdot x_1}$$

If x_0 and x_1 expressed from the latter equations are substituted into Eq. (1.30), the following relationship is obtained:

$$A = \frac{1}{2}kC \cdot \gamma(f_1 - f_0) = \frac{1}{2}kC \cdot \gamma \cdot \Delta f \qquad (1.31)$$

while the power required is:

$$P = \frac{A_1}{\Delta t} = \frac{1}{2}kC\frac{\Delta f}{\Delta t}G \qquad (1.32)$$

where Δt is the time for which the material is retained in the mill. The quotient $\Delta f / \Delta t$ expresses the mean rate of comminution, while the quotient $G/\Delta t$ gives the throughput Q per hour, and so Eq. (1.32) may be written in the form

$$P = \frac{\Delta A}{\Delta t} = \frac{1}{2}kC \cdot \Delta f \cdot Q \qquad (1.32a)$$

The equation may be developed further as follows:

$$P = \frac{1}{2}kC\frac{dF}{dt} = v\frac{dF}{dt} \qquad (1.33)$$

where dF/dt is the increment of surface area per hour in the mill, and v the specific energy of comminution (kWh/cm^2 or Ncm/cm^2).

1.7.2   General Relationships for Hammer Mills

In the stable operating state the quantity of material charged into the mill equals the quantity discharged. A certain quantity of material is always found in the grinding space, which is termed the "filling" (F). Thus if the quantities charged and discharged are plotted as function of time, then after the steady state of operation has been obtained two parallel straight lines are found (Fig. 1.33). The vertical distance between the two straight lines gives the magnitude of the filling F, while the horizontal section determines the grinding time t_g. The grinding capacity is defined by the simple relationship

$$Q = \frac{F}{t_g} \qquad (1.34)$$

from which the filling may be obtained as

$$F = Q \cdot t_g \qquad (1.34a)$$

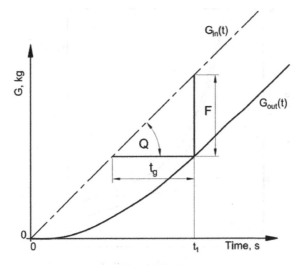

Fig. 1.33 Operation diagram for a hammer mill

The filling in the hammer mill increases with throughput and decreases with higher screen holes.

The governing laws and processing methods elaborated are also suitable to determine the performance of hammer mills for grinding wood materials. In the following we demonstrate this method for pine wood as an example. In a comminution process, the initial surface area of a particle will be increased (Δf, cm^2/g) and the increase in specific surface area consumes energy. The specific energy of comminution (v, kWh/cm^2) depends on several factors such as timber species and their mechanical properties, moisture content and slightly also on the geometric size distribution.

The net power consumption of comminution, without the idling power, is given by the simple equation as

$$P_e = v \cdot \Delta f \times 10^6 \text{ kWh/t}$$

There are several practical variables such as the

- geometric mean of particle size distribution,
- screen size used in the hammer mill (generally 1–10 mm)

which have strong interrelations with the increase in specific surface area Δf.

Figure 1.34 shows the relationship between the required surface increase and the geometric mean d_g supposing an initial mean particle diameter of 5 and 10 mm. This Figure also shows the specific energy of comminution calculated from the net power requirement.

The specific energy of comminution is roughly twice as high as those measured for maize and barley.

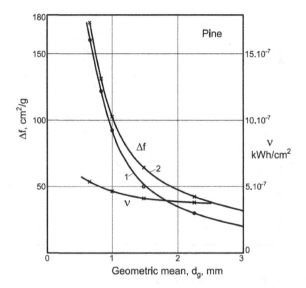

Fig.1.34 Increase in specific surface area and specific energy of comminution as a function of mean geometric diameter. Moisture content 12%. 1—$d_i = 5$ mm, 2—$d_i = 10$ mm

The curves in Fig. 1.34 can be described with the following simple empirical equations

$$\Delta f = 105 \cdot d_g^{-1.15}, \text{cm}^2/\text{g} \tag{1.35}$$

for the initial size of 5 mm.

Variation in the initial size slightly modifies the above equation

$$\Delta f = 91 \cdot d_g^{-1.3} \text{ cm}^2/\text{g} \tag{1.35a}$$

for $d_i = 10$ mm.

The specific energy of comminution is also the function of chip size

$$v = 4.8 \times 10^{-7} \cdot d_g^{-0.25} \text{ kWh/cm}^2$$

and combining the above equations yields:

$$P_e = v \cdot \Delta f = 50 \cdot d_g^{-1.4} \text{ kWh/t} \tag{1.36}$$

where the geometric mean d_g must be substituted in mm.

The screen size used in a hammer mill strongly influences the increase in surface area Δf and also the power consumption. Figure 1.35 shows the influence of screen size on the surface increase during comminution and the corresponding power consumption.

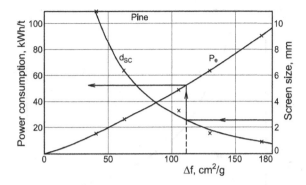

Fig. 1.35 Interrelation between increase in surface area, power consumption and screen size. Pine, moisture content 12%

The following equation holds

$$\Delta f = 164 \cdot d_{SC}^{-0.52} \; cm^2/g$$

which, combining with Eq. (1.35a), yields

$$d_g = 0.68 \cdot d_{SC}^{0.45} \; mm \tag{1.37}$$

where the screen size d_{SC} must be substituted in mm.

The moisture content of raw material has a significant effect on the specific energy of comminution and the specific power consumption. Quite similarly to crop products, the power consumption increases with the moisture content in a parabolic fashion and Eq. (1.36) can be supplemented in the following form

$$P_e = 50 \cdot d_g^{-1.4} \left[1 + 0.33(U - 12)^{0.5} \right] \; kWh/t \tag{1.38}$$

where the moisture content U (w.b.) is substituted in per cent.

It should be stressed that the numerical values given in the above equations refer to a pine wood with a density of around 500 kg/m³. Hardwoods require more energy for comminution compared to soft woods. Within one species, its density may have a definite effect on the required energy.

In practical cases more convenient variable is the geometric mean of the size distribution. Figure 1.34 gives a generally valid relationship between the specific surface area and the geometric mean. The net energy requirement is, however uniquely defined by the increment of the specific surface area. Figure 1.36 shows a direct relationship between the mean diameter of chips and the specific energy consumption. Below 1 mm mean diameter the energy consumption steeply increases.

If a more uniform particle size distribution is needed, the non-desirable fractions may be screened or to use the closed-circuit grinding method which may reduce the energy consumption in a certain extent (Henderson and Bölöni 1966).

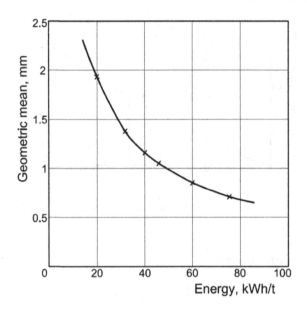

Fig. 1.36 Relationship between mean diameter of chips and specific energy requirement

Literature

Bölöni, I.: Some functional regularities of the comminution process in hammer mill. Acta. Tech. Acad. Sci. Hung. **41**, 381–398 (1962)

Bölöni, I.: Some regularities of grain distribution and fineness variation as observed in hammer mill products. Acta. Tech. Acad. Sci. Hung. **45**, 45–64 (1964a)

Bölöni, I.: The required power input of hammer mills. Acta. Tech. Acad. Sci. Hung. **45**, 327–344 (1964b)

Hamilton, G., Goodman, L.: The stress field created by a circular sliding contact. J. Appl. Mech. **33**, 371–376 (1966)

Henderson, S., Bölöni, I.: Closed-circuit grinding of agricultural products. J. Agric. Eng. Res. **11**, 248–254 (1966)

Kirbach, E.: Saw performance critical to lumber quality, yield. Wood Technol. **39**, 22–24 (1995)

Rittinger, P.: Lehrbuch der Aufbereitungskunde. Springer, Berlin (1867)

Sitkei, G.: Fortschritte in der Theorie des Spanens von Holz. Holztechnologie, No. 2, pp. 67–70 (1983)

Sitkei, G.: Mechanics of Agricultural Materials. Elsevier, Amsterdam (1986)

Sitkei, G.: On the mechanics of oblique cutting of wood. In: Proceedings of 13th IWMS Vancouver, pp. 469–476 (1997)

Sitkei, G., et al.: Theorie des Spanens von Holz. Fortschrittbericht No.1. Acta Fac. Ligniensis Sopron (1990)

Sugihara, H., et al.: Wood cutting with a pendulum dynamometer (VI). Wood Research No. 39, Kyoto, pp. 1–12 (1966)

Timoshenko, S., Goodier, J.: Theory of Elasticity. McGraw-Hill, New York (1951)

Chapter 2
Thermal Loading in Cutting Tools

2.1 Introduction

When wood is processed by machines, the chip slides on the surface of the tool and friction work transforms into heat. Most of this heat flows into the direct surroundings of the edge tool. Depending on the friction power, the area of the edge warms up and the temperature on the surface of the edge can reach a large value.

The high temperature of the edge of a tool increases the wear of a tool as heat weakens and softens metals—especially above 500–600 °C. Microscopic tests of the tools of the timber industry have shown that the smearing of the material is noticeable on the edge surface which is the consequence of the softening of the material. Recently deposits of calcium and potassium were observed on the knife surfaces which can be removed from the wood only at around 800 °C (Demyanovskij 1968; Stewart 1989).

Although the existence of high temperature on the edge surface is expected (Demyanovskij 1968; Stewart 1989), no direct measurements or calculations for determining the thermal state of the cutting edge have been presented yet.

In the following we give a detailed analysis of the heat generation in the contact surface, the rate of heat flow into the edge and rake face, the amount of heat divided into the edge and rake face, the amount of heat divided into the cutting edge and the wood and the heat transmission from a rotating tool into the surrounding air. Furthermore, the influence of tool geometry and operating conditions on the thermal loading of the cutting edge is presented. Working diagrams are constructed to facilitate the quick estimation of various influencing factors on the thermal state of the knife.

2.2 Heat Generation in the Contact Surfaces

Tools used in the metal and wood industry differ substantially from each other in one aspect. In cutting metals, the whole chip cross section deforms plastically and the resulting chip thickness is two to three times larger than the thickness of the chip that was set. This plastic deformation is associated with remarkable energy absorption and heat generation inside the chip.

In cutting wood, we cannot observe a similar occurrence and the cross section of the chip hardly change. Therefore the source of heat is the friction work done on the contact surfaces.

A pressure distribution develops on the edge and front surface of the knife (Fig. 2.1), while the chip moves with v speed in relation to the knife. The theoretical value of the generated heat is:

$$Q = v \int_F \mu\sigma \cdot df \ \text{(Watt)}$$

The pressure distribution on the knife is not known with sufficient accuracy, therefore we will use the basic relationships of the cutting theory.

The frictional power on the front surface of the knife is determined by the friction force and the sliding speed:

$$Q = \mu N \cdot v$$

where N is the normal force on the rake face.

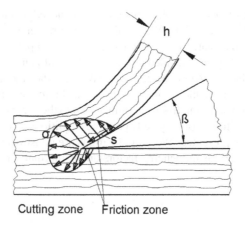

Cutting zone   Friction zone

Fig. 2.1 The stress distribution on the surface of the tool (Sitkei et al. 1990)

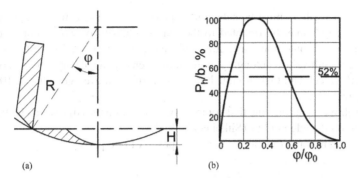

Fig. 2.2 The rotation angle of the tool during cutting (**a**), and the shape function of the cutting force (**b**)

Combining Eqs. (1.1), (1.7) and (a1.8) we may write:

$$N(\sin \delta + \mu \cdot \cos \delta) = B \cdot h$$

from which

$$N = \frac{B \cdot h}{(\sin \delta + \mu \cdot \cos \delta)} \cong B \cdot h \qquad (2.1)$$

as the value of the denominator is close to the unit in the range of $\delta = 60–70°$ and $\mu = 0.25–0.3$.

Rotating tools work only in a given rotation angle φ per turns (Fig. 2.2a), heat flows only in this range of angles. The value of the φ angle is calculated as:

$$\varphi = \arccos \cdot \left(1 - \frac{H}{R}\right)$$

Due to the changing cross-section of the chip, the cutting force varies in a similar way. Its average value related to the rotation angle of cutting way have different value. In order to get reliable average value, it is advisable to describe the force which can then be integrated for calculating the average value.

As we saw in the previous Chapter (see Eq. a1.8), the cutting force is linear as a function of chip thickness. For practical uses it is purposeful to express this relationship with the bending strength of the wood in the following form:

$$\frac{P_h}{b} = 4 + 0.3\sigma_b \cdot h \quad \text{N/mm} \ (\delta = 60°)$$

$$\frac{P_h}{b} = 5 + 0.4\sigma_b h \quad \text{N/mm} \ (\delta = 70°)$$

where the strength σ_b must be substituted in N/mm². The selection of appropriate bending strength may be oriented to the volume density of wood. For example, European Scotch pine may have quite different densities from 380 to 650 kg/m³. Bending strength values belonging to the above density range vary between 60 and 90 N/mm². Beech and Oak are much more uniform with average value of 105 and 110 N/mm².

In the above equations, the first component is acting on the edge which is strongly dependent on the edge radius (Sitkei et al. 1990).

The resultant force function may be composed of two parts

$$F\left(\frac{\varphi}{\varphi_0}\right) = F_{max} \cdot f\left(\frac{\varphi}{\varphi_0}\right)$$

where F_{max} means the maximum force during cutting and $f\left(\frac{\varphi}{\varphi_0}\right)$ is the shape function. The maximum force is given by the above force equations at the maximum chip thickness, while the shape function may be approximated by the following equation

$$f\left(\frac{\varphi}{\varphi_0}\right) = C\left(\frac{\varphi}{\varphi_0}\right)^n \cdot e^{-k\left(\frac{\varphi}{\varphi_0}\right)^m}$$

where the constants C, k, n and m are to be selected to fit the experimental force curve. The constant C adjusts the shape function to the unit height. With $R = 60$ mm, $e_z = 1$ mm, $H = 4$ mm, the cutting angle is $\varphi_0 = 21°$, the maximum force for beach wood was 29 N/mm and the best fit was obtained for $C = 2.5$, $n = 0.5$, $k = 6$ and $m = 3$. The average force integrated along the cutting angle is 52% of maximum value, which is 15.08 N/mm distributed along the arc length $R\varphi_0 = 22$ mm.

Figure 2.2b shows the shape function in relative units.

The derived calculation method greatly facilitates the correct estimation of the average cutting force required in energy consumption and friction energy calculations. The latter is the most important input data for determining the thermal load on tool edges (Sitkei et al. 1990; Csanády 1993).

The ratio of the equivalent cutting force, as an average, related to the cutting arc length and the maximum force gives the effective force ratio

$$\psi = \frac{P_{he}}{P_{h\,max}}$$

Usually the value of Ψ varies between 0.5 and 0.6. The total heat introduced on the rake face is:

$$Q = \mu Bhbv\psi\left(\frac{\varphi}{360}\right)$$

The equivalent length of the heat inflow on the rake face depends primarily on the thickness of the chip and its value, from experimental observations and theoretical considerations, may be taken as:

$$L = (1.2 \sim 1.3) \cdot h$$

The theoretical value of the specific heat flow on the rake face can be written as follows:

$$q = \tfrac{1}{1.3}\mu B b v \psi \left(\tfrac{\varphi}{360}\right) \left[\tfrac{W}{mm}\right] \tag{2.2}$$

Taking into consideration the expression of the force acting on the edge (see Eq. a1.8), the theoretical value of the specific heat flux to the edge will be:

$$q_a = \mu \sigma_c b v \psi_1 \left(\tfrac{\varphi}{360}\right) \left[\tfrac{W}{mm}\right] \tag{2.3}$$

The arc length of the edge:

$$L_a = \rho(\pi - \delta)$$

where

ρ means the radius of the edge
δ is the cutting angle and its value has to be substituted in radian
ψ parameter takes into account a force decrease at the exit point of the knife with $\psi_1 = 0.7$ value.

Figure 2.3 demonstrates the change of the specific heat fluxes depending on the H/R ratio ($v = 37.7$ m/s). As we can see, the heat flux introduced on the edge is more than double compared to the heat on the rake face.

2.3   Boundary Conditions

The temperature field developing in the knife depends on many factors. The main factors are the following:

(a) the geometric characteristics of the knife, the heat conductivity of the knife material and its specific heat,
(b) the true heat flow is smaller than the theoretical value both on the edge and on the rake face, because friction is an irreversible process. Furthermore, friction, as a shearing stress, generates deformations in both contacting materials, which also consumes energy, and finally some heat will be dissipated into the air,
(c) intensive heat transfer occurs on the surface of fast rotating tools that partly decreases the amount of heat and cools the knife while rotating freely in the air,

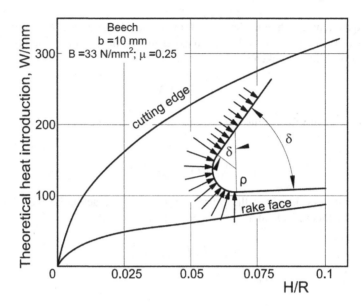

Fig. 2.3 The theoretical distribution of the heat flow on the rake face and on the edge of the tool ($v = 37.7$ m/s) (Sikei et al. 1990; Csanády 1993)

(d) the frictional heat is generated on the contact surface of the chip and the tool, so a smaller part of the heat flows into the chip, the greater part flows into the knife,

(e) a further possible inaccuracy may be given by the fact that the effective friction coefficient between the tool and wood cannot be measured during the cutting process. Friction coefficient values available were determined from cutting force measurements.

The previous questions can be answered partly from experimental results and partly from theoretical considerations. If the boundary conditions (the location of the heat flow and intensity, heat transfer on the surface) are known, then the differential equation of the heat conduction can be solved with a computer (finite element method) and the whole temperature field can be determined.

The comparison of temperature measurements made inside of the knife and the theoretical calculations showed that 92–95% of the theoretically determined frictional heat appears as effective heat on the frictional surface. In the following calculations, we have taken 95% of the theoretical values calculated with Eqs. (2.2) and (2.3).

The smaller portion of the generated heat is taken away by the chip. This amount of heat can be estimated as follows. The temperature variation of the infinite half-space applying a constant surface heat flux can be calculated from the following equation (Carslaw and Jaeger 1959):

$$Q\vartheta = \frac{q_0}{\lambda}\sqrt{\frac{4at}{\pi}}\left[e^{-x^2/4at} - x\sqrt{\frac{\pi}{4at}}\cdot erfc\left(\frac{x}{\sqrt{4at}}\right)\right]$$

where:

$$erfc = 1 - \frac{2}{\sqrt{\pi}}\int_0^z e^{-\beta^2}dz$$

λ heat conduction coefficient,
q_0 heat flux on the surface,
t the time,
a temperature conductivity.

The temperature of both materials is the same at the contact surface; therefore, the following equality can be written for the contacting surfaces ($x = 0$):

$$\frac{q_w}{\lambda_w}\sqrt{\frac{4a_wt}{\pi}} = \frac{q_k}{\lambda_k}\sqrt{\frac{4a_kt}{\pi}}$$

from which, the heat ratio is calculated with the following equation:

$$\frac{q_w}{q_k} = \frac{\lambda_w}{\lambda_k}\sqrt{\frac{a_k}{a_w}} \tag{2.4}$$

In the previous equations, the w index is related to the wood, and the k index to the knife. The heat conductivity and the temperature conductivity coefficients of the wood depend on its moisture content. The values of λ and a are summarized in Table 2.1.

Calculations showed that 3.5–10% of the total heat flows into the chip depending on its moisture content.

This is in agreement with practical observations that cutting wet wood considerably decreases the knife-temperature.

When machining wood with high moisture content, water is pressed out to the surface and the heat needed for evaporation further decreases the surface temperature.

Table 2.1 Thermal parameters

Material	λ, W/m°C		a, m²/h	
	U = 0%	U = 30%	U = 0%	U = 30%
Wood	0.163	0.36	0.00055	0.00042
Steel	40		0.042	

Note U is the moisture content, wet basis

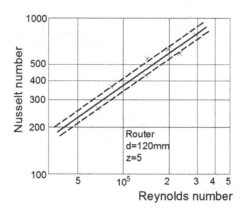

Fig. 2.4 Similarity relationship of the heat transfer criteria for milling tools (Csanády 1993)

Most of the heat amount flowing into the tool radiates from the surface of the tool into the air. A complicated convective heat transfer develops around the tool, so the heat transfer coefficient can only be reliably determined with experiments. The method is the following: after the tool has reached a steady heat condition, cutting is interrupted and the cooling down curve (see Fig. 2.12.) is registered at some points of the rotating tool. Then we calculate the cooling down curve of the point, (finite element method), by using different heat transfer coefficients. When the measured and the calculated cooling down curve coincide with each other, the heat transfer coefficient agrees the real value. The results of these kinds of measurements and calculations are demonstrated in Fig. 2.4. The result can be described with the general similarity equation (Csanády 1993)

$$Nu = 0.12 \cdot Re^{0.7} \tag{2.5}$$

with

$$Nu = \frac{\alpha d}{\lambda} \text{ and } Re = \frac{v \cdot d}{\upsilon}$$

where:

Nu   Nusselt number,
Re   Reynolds number,
v circumferential speed of the tool,
d diameter of the tool,
α heat transfer coefficient,
λ the heat conductivity coefficient of the air,
υ the kinematic viscosity of the air.

It is interesting to note that the similarity equation valid for the plane wall ($Nu = 0.032\ Re^{0.8}$) gives somewhat similar results.

2.4 The Temperature of the Tool in Different Operating Conditions

Concerning thermal condition of a tool, the following characteristic operating conditions may be distinguished:

(a) true steady state operating (e.g. wood turning), when the temperature remains constant, independently of time at each point of the tool;
(b) quasi-steady state operating (rotating tools), when the temperature of the tool remains nearly constant, except for the surface and a layer a few tenths of millimetres under it;
(c) unsteady state (not stationary) operating condition at the heat introduction, on and under the surface in a thin layer (in rotating tools see Fig. 2.5);
(d) unsteady state in the whole volume of the tool, the warming up and cooling down period of the tool.

Rotating tools always work in an unsteady heat state. The duration of the inflow of heat is just $\varphi = 15\text{--}25°$ per turn. During the remaining part of the angle of rotation, the surface of the tool edge cools down (Fig. 2.5). Therefore on the surface of the edge the temperature is highly variable around the steady state value.

The amplitude of the temperature variation decreases rapidly as a function of depth. If we approximate the surface temperature variation with the harmonic function

$$\vartheta = \vartheta_0 \cos\left(\frac{2 \cdot \pi \cdot n \cdot t}{t_0}\right)$$

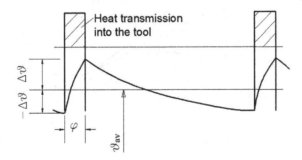

Fig. 2.5 Periodic heat inflow and temperature variation in rotating tools (Sitkei et al. 1990)

Then the temperature variation under the surface, depending on the depth and the time can be calculated from the following expression (Carslaw and Jaeger 1959):

$$\vartheta_{x,t} = \vartheta_0 e^{-\sqrt{K} \cdot x} \cos\left(\sqrt{K}x - \frac{2 \cdot \pi \cdot n \cdot t}{t_0}\right),$$

where:

$$K = n\pi/at_0$$

and

n $1, 2, 3, \ldots,$
t_0 the time of one revolution,
a the temperature conductivity.

The amplitude will be the greatest, if the cos-term value is one. In this case the decrease of the temperature amplitude as a function of depth (Fig. 2.6) will be calculated as

$$\frac{\vartheta_{x,t}}{\vartheta_0} = e^{-\sqrt{K} \cdot x}$$

or

$$x = \sqrt{K} \cdot \ln\left(\frac{\vartheta_0}{\vartheta_{x,t}}\right) \tag{2.6}$$

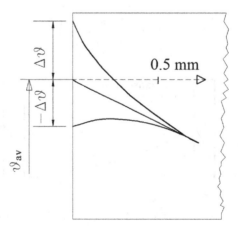

Fig. 2.6 Decrease of the temperature amplitude depending on the depth

Taking a rotation speed of 6000 rpm, for which $t_0 = 0.01$ s, to a 1/10 or 1/100 amplitude ratio will correspond to $x = 0.44$ mm or 0.88 mm depth, respectively. Due to the short heat inflow, the real x values are smaller, as the calculations with the finite element method have shown. We concluded that the cyclic fluctuation of the temperature around the site of heat inflow can be neglected at a depth of 0.5 mm.

Some problems may arise when calculated and measured temperatures are compared within the body of the tool. In theoretical calculations it is supposed that in steady state operation the heat inflow is in equilibrium with the heat carried away by conduction and surface heat transfer into the air. In this case the middle part of the tool is never overheated and its temperature is slightly over the ambient temperature. In practice, however, it can happen that the heat remove is hindered and the temperature of the middle part increases in order to be in equilibrium. In this case the temperature field will be more equalized with smaller temperature gradients. This principle can be utilized to decrease temperature differences along the saw blade radius by either local heating or cooling to achieve blade stability (Fig. 6.24).

The steady temperature of rotating tools primarily depends on the thickness of the chip, the edge rounding radius, the sharpening angle β and the peripheral speed. Less influencing factors are the strength properties of the wood and the moisture content, the arc length of the cutting, the overload of the gullet with sawdust etc.

The calculations with the finite element method showed that the maximum temperature always occurs on the surface of the edge, approximately at the intersection point of the bisector and the circumference of the edge. Next we review the basic relationships of this maximum temperature as a function of the influencing factors. All measurements and calculations refer HSS steel.

Figure 2.7 shows the variation of the maximum temperature depending on the rotation speed ($d = 120$ mm). The parameter x notates the heat inflow length on the rake face, which is proportional to the thickness of the chip ($x = (1.2 \sim 1.3)h$). The maximum temperature increases by the 0.8 exponent of the rotation speed or peripheral speed.

Figure 2.8 shows the maximum temperature depending on the heat inflow length for different edge rounding radii (different sharpness). As we can see, the edge wear substantially increases the maximum temperature.

Figure 2.9 shows measured and calculated maximum temperatures as a function of feed per tooth (Csanády 1993; Sitkei and Csanády 1994). Measurements were taken in the bisector of the edge and a thin thermocouple was located at $\Delta x = 0.7$ mm from the edge surface. The empirical equation describing the temperature variation has the form:

$$\vartheta_x = \vartheta_0 + [180 + 465 e_z^{0.5}] e^{-0.3\Delta x}$$

where ϑ_0 means the ambient temperature. Obviously on the surface $\Delta x = 0$ holds.

The sharpening angle β decisively influences the conduction of heat to the air; therefore, the temperature depends strongly on the β angle. In Fig. 2.10 we can see the

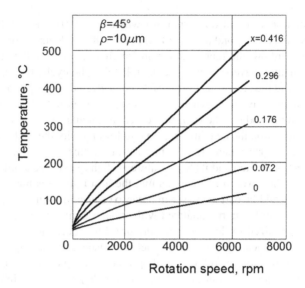

Fig. 2.7 The maximum temperature of the edge surface depending on the rotation speed for different length of heat inflow on the rake face (x in mm) (Sitkei et al. 1990, Sitkei and Csanády 1994)

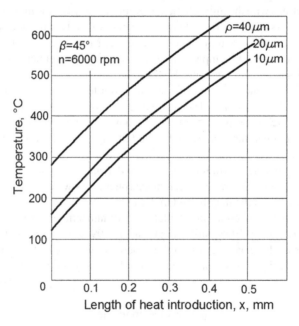

Fig. 2.8 The effects of edge wear on the maximum temperature for different length of heat inflow (Sitkei et al. 1990; Sitkei and Csanády 1994)

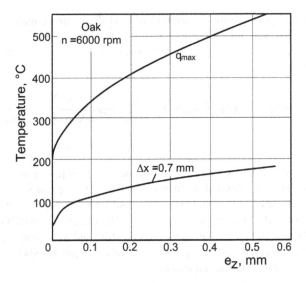

Fig. 2.9 Measured and calculated maximum surface temperatures and temperatures under the edge as a function of feed per tooth (Csanády 1993; Sitkei and Csanády 1994)

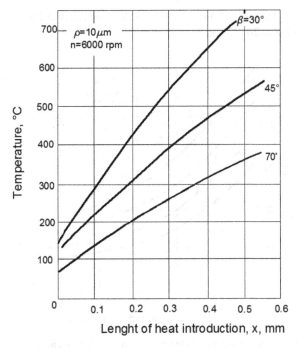

Fig. 2.10 The effect of a sharpening angle on the maximum edge temperature (Csanády 1993; Sitkei and Csanády 1994)

variation of the maximum temperature for different sharpening angles as a function of the length of the heat inflow.

The curves of the Figs. 2.7, 2.8 and 2.10 can be described with the following general equation:

$$\vartheta_{max} = \vartheta_0 + \left[26.6\rho^{0.8} + 1503 \cdot x^{0.8}\right] \frac{v^{0.8}}{\beta^{0.95}\left[\text{unknown template}\right]} \qquad (2.7)$$

where:

ϑ_0—is the ambient temperature and the value of ρ is in μm, x is in mm, v is in m/s and β should be substituted in degrees. In Fig. 2.11 we can see experimental results of circular saws ($d = 400$–500 mm) measured at the bisector of the teeth at a distance Δx from the edge (Sitkei et al. 1990). The extrapolation of the measured points until $\Delta x = 0$ gives the edge surface temperature and these temperature values can be described well with Eq. (2.7). The temperature of any points on the bisector can be given with the following equation:

$$\vartheta_x = \vartheta_{max} \cdot e^{-k \cdot \Delta x^n}$$

Using the data from Fig. 2.11, we get the following relationship. ($\beta = 55°$ and $\rho = 20$–22 μm):

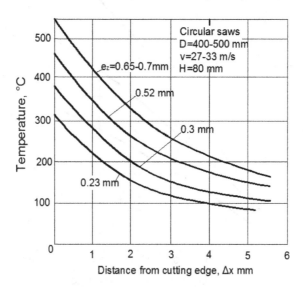

Fig. 2.11 Variation of the temperature in the bisector of the edge for circular saws working with different feeds per tooth (Sitkei et al. 1990)

$$\vartheta_x = \vartheta_0 + \left[7 + 586\left(\frac{\varphi}{360}\right)e_{zav}^{0.8}\right] \cdot v^{0.8} \cdot e^{-0.4\Delta x^{0.65}} \qquad (2.8)$$

where e_{zav} means the average feed per tooth in mm.

The typical operating condition of cutting tools is the warming up and the cooling down periods. In Fig. 2.12 we can see typical warming up and cooling down curves for a milling cutter at five different points located on the bisector (Csanády 1993; Sitkei and Csanády 1994). Both the warming up and the cooling down are relatively fast, the time constant chiefly depends on the distance from the edge. (Time constant is defined as the time required to reach 63.2% of the maximum temperature increase) (Fig. 2.13).

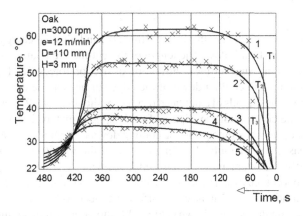

Fig. 2.12 The warming up and cooling down of a tool measured in the bisector and in different distances from the edge surface (Csanády 1993)

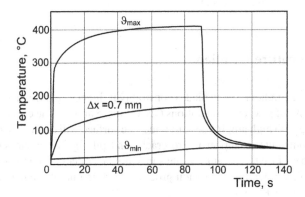

Fig. 2.13 Calculated warming up curves at different points of a tool (Csanády 1993; Sitkei and Csanády 1994)

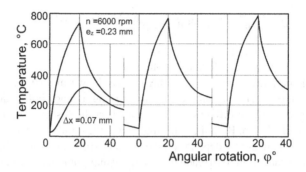

Fig. 2.14 The first three cycle of an unsteady warming up process as a function of angular rotation (Csanády 1993; Sitkei and Csanády 1994)

Using circular saws, the length of the cutting arc, and also the energy of friction forces influences the heat inflow. The length of the cutting arc related to the cutting height ($\varphi R/H$) is a measure for the contribution of friction forces (see Fig. 3.11).

The warming up and cooling down of the edge surface can be determined only with calculation. These kinds of calculated results are shown in Fig. 2.14 as a function of time. The time constant of the edge surface is approximately 2 s. Moving away from the surface, the warming up and the cooling down take more time, and the time constant increases considerably. The temperature of the tool itself changes slowly and only slightly at 3–4 cm away from the edge surface.

Rotating tools cut only for a short time within one revolution. Therefore a cyclical temperature variation occurs on the edge surface that decreases quickly as a function of depth. Those calculations are shown in Fig. 2.14, showing the temperature variation of the first three cycles depending on the angle rotation. The amplitudes are very large, and after reaching the steady state temperature (400–500 °C) the peak temperature can spike to 1000 °C for an instant.

2.5 Temperature Gradient

The temperature gradient in saw blades considrably influences the frequency of vibration and stability of saw under operations (see Chaps. 5 and 6). The cyclic fluctuation of temperature with high amplitudes just under the surface will cause thermal stresses which may contribute to appear micro cracks and pitting of the surface. Furthermore, the high surface temperature softens in a thin layer decreasing the strength and may cause high temperature corrosion process which is one of the majot tool wear mechanism (see Sect. 7.3.).

The temperature gradient in the tool body depends on tool geometry and operating parameters. The most important parameters of tool geometry are

- the edge radius,
- the sharpening angle,
- the tool diameter.

The major operating parameters are

- the rotation speed,
- the feed per tooth or length of heat introduction,
- the depth of cut.

The most important material properties are

- the wood density,
- the direction of cut,
- resin content and surface coating,
- the moisture content.

The tool diameter and rotation speed have a combined effect. They determine the cutting speed which governs the specific heat flux to both the rake face and the edge (see Eqs. 2.2 and 2.3).

It is important to mention that the temperature gradient is the highest just under the surface and generally decreases as a function of depth. As an example, Fig. 2.15 shows the variation of temperature gradient.calculated with FEM ($\beta = 45°$, $\rho = 20\ \mu m$, $v = 37.7\ m/s$) as a function of depth. Just under the surface a temperature gradient of 6000 °C/mm is a typical value. At a depth of 0.1 mm the temperature gradient is roughly halved compared to the highest value.

The effect of edge radius on the temperature gradient shows an interesting picture. Our FEM calculations has shown that, decreasing the edge radius, the surface temperature and its gradient continuously increases. At higher edge radii the heat flux on the edge increases but the cross section for heat conduction will also be higher. This combined effects result in a more or less constant temperature gradient. Figure 2.16.

The heat flux is directly proportional with the cutting speed and, therefore, the rotation speed for a given tool influences the temperature gradient in the same manner, Fig. 2.17. That means that the cutting speed is a decisive operation parameter in every respect.

Especially the tool life between two sharpening fundamentally depends on the cutting speed (see Sect. 7.3).

The effect of chip thickness or length of heat introduction shows also an interesting feature. A bigger chip thickness means higher heat introduction but also a higher length of introduction. The local heat load, therefore, remains nearly constant and the temperature gradient varies not much with chip thickness.

The sharpening angle has a definite influence an the temperature gradient. Using lesser sharpening angle, the cross-section for heat conduction decreases which results in a higher temperature gradient, Fig. 2.18.

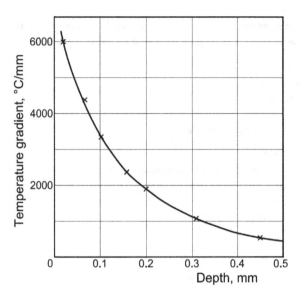

Fig. 2.15. Temperature gradient as a function of depth. $D = 120$ mm, $n = 6000$ rpm, $\beta = 45°$, $\rho = 20\ \mu$m

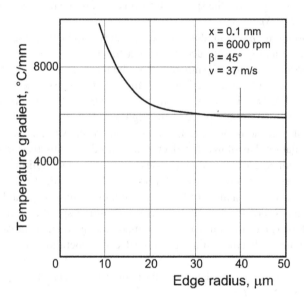

Fig. 2.16 Maximum temperature gradient as a function of the edge radius

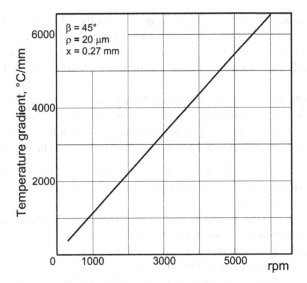

Fig. 2.17 Maximum temperature gradient as a function of rotational speed. $D = 120$ mm, $v = 37$ m/s

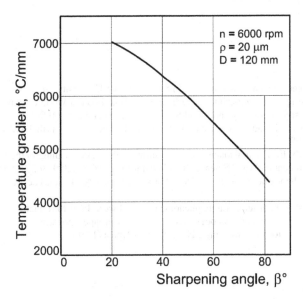

Fig. 2.18 Relationship between maximum temperature gradient and sharpening angle

Finally it should be noted that the shape of tool has also some effect on the temperature field. Thin-wall constructions, such as saw baldes, have a greater surface-volume ratio and, therefore, the cooling into the air is more effective. Casing of a tool may also modify the cooling conditions and hereby the temperature field in certain extent.

In recent time the high speed machining became a common practice. Due to the high thermal loading of the tools, the proper selection of edge materials or using special edge coatings is one of the most important task in the wood industry.

From the previous discussion the following conclusions can be drawn:

- the heat load in woodworking tools primarily originates from frictional heat;
- the high peripheral speed generates large frictional power and, as a consequence, high surface temperatures;
- the high surface temperature is an important cause of wear, as the materials soften and high temperature corrosion occurs (see in Chap. 7);
- the high cyclic temperature variations of the edge surface causes cyclical heat-stress that can cause pitting of the surface;
- the edge temperature of a worn tool is always larger than for a sharp tool which further increases the intensity of the wear.

Literature

Carslaw, H., Jaeger, J.: Conduction of Heat in Solids. Oxford University Press, New York (1959)

Csanády, E.: Faforgácsoló szerszámok hőterhelése (Thermal Load in Wood Cutting Tools). Ph.D. Thesis, University of Forestry and Wood Sciences, Sopron, Hungary (1993)

Csanády, E.: Thermal load in wood cutting tools. In: Proceedings of 17th IWMS, Rosenheim, pp. 28–43 (2005)

Sitkei, G., Csanády, E.: Thermal loading in wood cutting tools. In: Proceedings of the 1st Conference on the Development on Wood Science, Buckinghamshire, UK, pp. 359–368 (1994)

Sitkei, G. et. al.: Theorie des Spanens von Holz. Fortschrittbericht No.1. Acta Fac. Ligniensis Sopron (1990)

Stewart, H.: Feasible high-temperature phenomena in tool wear. Forest Product J., 25–28 (1989)

Демяновский, К.: Износостойкость инструмента для фрезерования древесины, (Wear resistance of tools for woodworking). Изд, Лесная Пром (1968)

Chapter 3
Operating Parameters of Wood Cutting Tools

3.1 Introduction

It is important to operate woodworking machines and tools economically and correctly. The optimal operating parameters and manufacturing costs change depending on the task. The operational parameters also have limits. Exceeding these limits may hamper the safe operation of the machine or the tool.

The most important operating parameters are the following:

- the peripheral speed of the tool,
- the feed speed,
- the feed per tooth,
- the depth of cut.

The operational parameters influence the following for a given tool:

- the kerf volume cut in the unit time,
- the cross section cut in the unit time,
- the surface roughness and waviness,
- the wear of the tool,
- the energy consumption of cutting,
- the costs of machining.

The ultimate goal of any woodworking operation should be to minimize energy consumption and manufacturing costs, and to maximize throughput and quality.

In the next section, we give the most important relationships of cutting tools used in woodworking operations.

3.2   The Bandsaw

The bandsaw has a straight and uniform motion during work, while the material is pushed against the saw perpendicular to the blade. The kinematic relations and notations are shown in Fig. 3.1. The parameters of the saw blade are its thickness (s), the extent of its spring set or the swage set ($2a$), the volume of its gullet (V_g) and the pitch of its teeth (t).

The speed of the cutting is practically the same as the band speed, as the feed speed is relatively small. The time needed to pass the length of the tooth-pitch is

$$\tau = \frac{t}{v}$$

during this time the material is pushed forth with e_z value,

$$\tau = \frac{e_z}{e}$$

and the feed per tooth will be

$$e_z = \frac{e}{v} t. \tag{3.1}$$

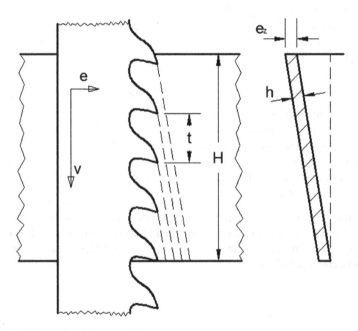

Fig. 3.1   The kinematic relations of the bandsaw

The v speed of the tool is the same as the peripheral speed of the band mill wheel

$$v = \frac{D \cdot \pi \cdot n}{60}$$

where: D is the wheel diameter and n is the rotation speed.

The solid wood volume cut by one tooth with $b = s + 2a$ width is calculated by

$$V = l \cdot h \cdot b = e_z H \cdot b = \frac{e}{v} H \cdot t \cdot b$$

where:

l the length of chip,
H the height of the cut cross section.

The chip and solid wood volume ratio is generally $\varepsilon = 3\text{--}4$ related to the solid wood, and with this the loose chip volume due to woodworking is:

$$V_{chip} = \varepsilon \frac{e}{v} H \cdot t \cdot b \tag{3.2}$$

The chip has to fit in the gullet until its exit from the material, therefore $V_g = V_{chip}$. Therefore the value of the maximum theoretical feed speed is from Eq. (3.2):

$$e_{theo} = \frac{V_g \cdot v}{\varepsilon \cdot H \cdot t \cdot b} \tag{3.3}$$

where V_g means the gullet volume.

We can define the *gullet feed index* as the ratio of the loose chip volume and the gullet volume (gullet area x kerf width).

The measurement of compaction of the chip has shown that only 0.2–0.4 bar pressure is needed for a 20–25% compaction. As a consequence, in practice the theoretical value of the feed speed can be increased by 25%. Figure 3.2 shows the mechanism of compaction and the development of the lateral pressure in the gullet. The lateral pressure coefficient changes between 0.25 and 0.35 in static pile of chips. Due to the movement of chips in the gullet, greater values are expected in the gullet than in the static case. If a greater amount of chips are forced into the gullet than the allowable, then the increased lateral pressure generates large frictional forces, raising the temperature of the tooth while the cutting force increases.

Figure 3.3 shows compaction measurements on saw chips in dry and wet conditions (Sitkei et al. 1990). The corresponding strain values are also given.

The band saw can be spring set or swage set. A swage set tooth makes a better surface and its cutting width is smaller compared to the spring set. The stability of a swage set tool is also better and, therefore the sawing deviation of long straight cuts is much smaller.

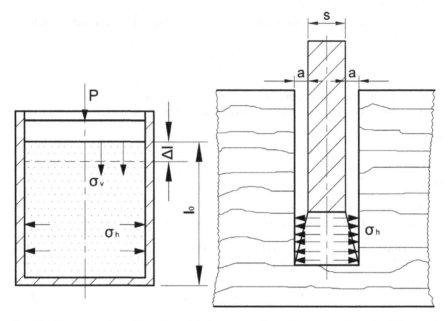

Fig. 3.2 The compaction of the chip and the formation of lateral pressure in the gullet (Sitkei et al. 1990)

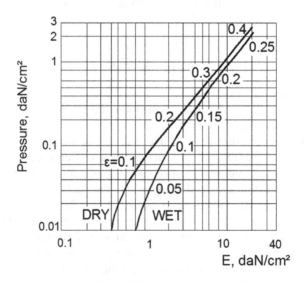

Fig. 3.3 Compaction of saw chips in dry and wet conditions, showing the corresponding strain values

3.3   The Frame Saw

The frame saw has a reciprocating movement and, because of the finite length of the connecting rod, the movement differs from the harmonic motion. The vertical displacement of the frame can be calculated from the following equation depending on the rotation angle (Fig. 3.4):

$$x = r\left(1 - \cos\varphi + \frac{\lambda}{2}\sin 2\varphi\right) \tag{3.4}$$

where:

r crank radius,
φ rotation angle,
l the length of the connecting rod,
$\lambda = r/l$ the ratio of the crank radius and the length of the connecting rod.

The continuous feed of the material can be possible in that way that the frame is inclined forward with κ angle, so by the moving upwards of the frame it makes a way free in horizontal direction:

$$y_1 = x \cdot tg\kappa$$

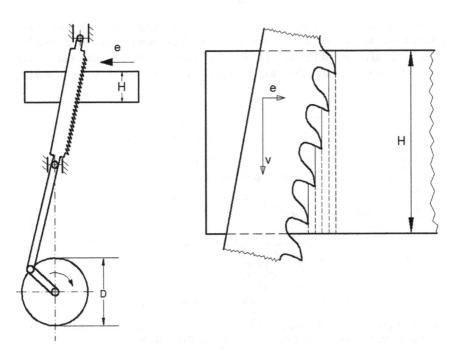

Fig. 3.4 The kinematic relations of the frame saw

The feed has a constant speed, and the corresponding displacement is

$$y_2 = e \cdot \tau$$

It follows that there is a close relation between the inclination of the frame and the feed speed. The following equations can be written for a half turn ($\varphi = 180°$):

$$\tau = \frac{\varphi}{6n} = \frac{30}{n} \quad \text{and} \quad \tau = \frac{y_1}{e} = \frac{2r \cdot tg\kappa}{e};$$

and equating the above two equations yields

$$e = \frac{2r \cdot n \cdot tg\kappa}{30} \tag{3.5}$$

The feed in one turn ($\varphi = 360°$)

$$e_n = e \cdot t = 4r \cdot tg\kappa \tag{3.6}$$

The average value of the feed per tooth is given by

$$e_z = e_n \frac{t}{2r} = 2t \cdot tg\kappa \tag{3.7}$$

Figure 3.5 (indicator diagram) represents the kinematic relations of the frame saw. If we change the inclination under the same conditions, then the sawing begins after the upper dead point at different angle. It changes the effective feed per tooth and the extent of the overfeed after the lower dead point. The overfeed phenomenon always

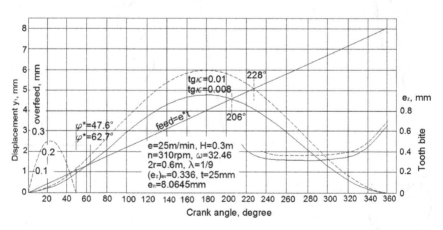

Fig. 3.5 Working diagram of a frame saw for two different forward inclinations, variation of tooth bite and overfeed as a function of the angle of rotation (Sitkei et al. 1990)

occurs because the movement of the frame is slow around the dead point, while the feed speed is constant.

One of the characteristic features of the frame saw is the dynamics of logs passing the bottom dead centre. This problem is of kinematic nature, i.e. the frame stops at the bottom dead point, while the feed of the log remains continuous. As a consequence, the log will be pushed onto the saw blades producing dynamic load in the blade and locally on the edges. This dynamic load depends on the feed rate, the pitch of teeth and on the forward inclination of the frame.

In order to reduce these dynamic loads, some manufacturers offer their machines with a feed interrupting mechanism built into the hydraulic drive of the feed rolls. However, the interrupting mechanism must be accurately adjusted; otherwise there is a considerable feed loss.

The tooth tip trajectories are needed in order to determine the overfeed values. The path of a tooth tip within the log can be calculated and plotted as shown in Fig. 3.6 (Sitkei and Horváth 1995).

The overfeed values depend on the pitch of the teeth and the feed per revolution. This relationship is given in Fig. 3.7. The overfeed values are relatively large and, therefore, the back face of the tooth cannot deform the wood in a measure of the overfeed value. As a consequence, the log must be stopped at the bottom dead centre. Stopping the log will cause an increased slip between the feed rollers and the log.

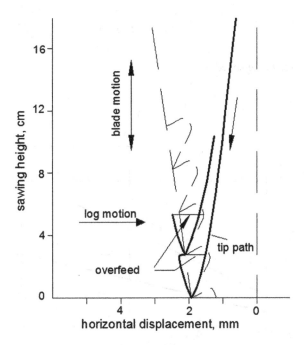

Fig. 3.6 Tooth tip trajectories to determine overfeed values

The above statement was experimentally verified. Figure 3.8 shows the log motion as a function of crank angle. Just after the bottom dead centre, the log stops and this interval is 50–60° of the angular rotation. That means that the pushing force is not enough to deform the wood to an extent which would be equal to the overfeed value (Sitkei and Horváth 1995).

The measurement of maximum pulling forces at the feed rolls gave values between 7000 and 8500 N. The greater values are valid for dry logs without bark. With 6–8 blades, this force exerts a load per tooth of around 150 N. Some additional loads originate from the deceleration of the log. This load per tooth is 40–60 N.

The slip between feed rolls and the log depends on many factors such as the wear of ribs and thorns of the feed rolls, the feed speed, the dullness of the saw teeth, the log surface (bark) and its moisture content.

During upward motion of the blades, no counter-forces act on the log (except a short distance near the bottom dead centre) and in this case the slip is the smallest possible and varies around 10% (Fig. 3.9). During cutting the slip increases due to cutting forces acting as counter-forces. The total slip related to a complete revolution includes the stopping of the log at the bottom dead centre.

The chip has to fit in the gullet when using frame saw. The chip volume can be calculated similarly, as with a band saw but we have to take into consideration that the frame saw cuts only per half-turns. As a consequence, the chip volume amounts to:

$$V_{chip} = 2 \cdot \varepsilon \frac{e}{v_k} H \cdot t \cdot b,$$

where the average blade speed is

$$v_k = \frac{D \cdot n}{30}$$

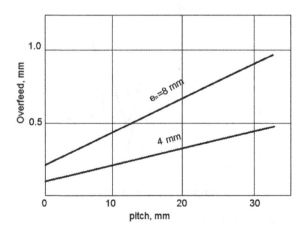

Fig. 3.7 Overfeed values as a function of pitch and feed per revolution

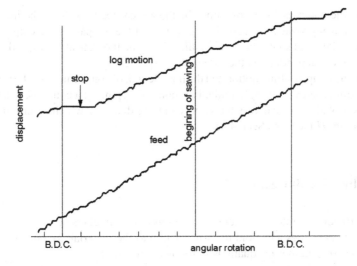

Fig. 3.8 Log motion as a function of crank angle with continuous feed

With this the theoretical feed speed is:

$$e_{th} = \frac{V_g \cdot v_k}{2\varepsilon \cdot H \cdot t \cdot b}$$ (3.8)

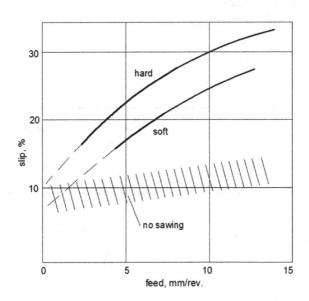

Fig. 3.9 Slip variation between feed rolls and logs for soft and hard woods (Sitkei and Horváth 1995)

Due to the compressibility of chips, the theoretical feed speed can be increased by 25–30% using swage set teeth. Using spring set teeth a part of the chip can get out of the gullet, and so it relieves the gullet. Then the theoretical feed speed can be nearly doubled, depending on the extent of the spring set.

Due to the reciprocating motion and the finite length of connecting rod, the rotation speed of frame saws is strongly limited to round 320 rpm. As a consequence, using a given sawblade, the real variable is only the log diameter which determines the cutting height H (see also Sect. 4.4).

3.4 The Circular Saw

Figure 3.10 demonstrates the kinematic relations of the circular saw. The tooth of the circular saw describes a cycloid track in relation to the material. The displacement is described by parametric equations in following manner:

$$x = R \cdot \sin \omega \tau + e \cdot \tau$$
$$y = R \cdot \cos \omega \tau$$

where:

τ the time,
R the radius of the circular saw,
ω the angular speed.

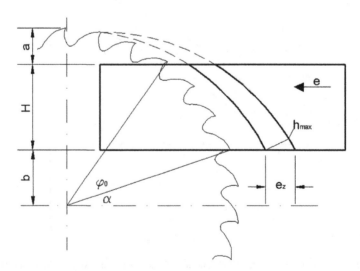

Fig. 3.10 The kinematic relations of the circular saw

The value of the feed per tooth is

$$e_z = \frac{e}{n \cdot z}$$

where:

z is the number of teeth.

The maximum value of chip thickness is approximately

$$h_{max} = e_z \cdot \cos \alpha$$

The average chip thickness depends on the H height and the position of the material. For the same H cutting height the φ_0 cutting rotation angle will be larger if the material is placed higher over the centre of rotation (distance b in Fig. 3.10).

The average chip thickness is

$$e_{zav} = \frac{e_z \cdot H}{\varphi_0 R} \qquad (3.9)$$

where the angle of the cutting φ_0 must be substituted in radian. The angle φ_0 can be calculated as follows

$$\varphi_o = \arcsin\left(\frac{b}{R} + \frac{H}{R}\right) - \arcsin\left(\frac{b}{R}\right)$$

It can be stated that the angle φ_0 depends on the position of the workpiece in relation to the top of the saw blade. The relative cutting arc influences the friction work of the sawing process. Therefore, the use of a lower machine table is preferable from energetic point of view.

The relative cutting arc is given by the following equation:

$$L_c = \frac{\varphi_0 R}{H}$$

which is illustrated in Fig. 3.11 for different machine table positions as a function of relative cutting height.

It is important to minimize the breaking of the edges of the workpiece (splintering). The direction of cutting forces has a great influence on edge quality. Compression forces provide a better cutting process without edge breaking; therefore, tension forces near the edge should be avoided. Edge breaking can be minimized for the upper or lower edge or both. According to experiments determined that the optimum overhang of the blade rim over the workpiece is around 10 mm as shown in Fig. 3.12 (Westkämper and Freytag 1991).

Circular saws are usually very noisy. Its noise can be reduced with the decreasing of the speed and the feed per tooth, and by increasing of the number of working teeth. It is also effective to coat the surface of the tool with cushioning material, but this raises the cost of the tool.

Cutting precious timber into thin lamellae (top layer for strip flooring parquet, pencil manufacturing), the kerf losses should be minimized using thin-kerf sawing. They require careful preparation after each sharpening to avoid instability (see Sect. 5.3). Using sawblades of 400–500 dia., the attainable minimum width is between 1.0 and 1.2 mm.

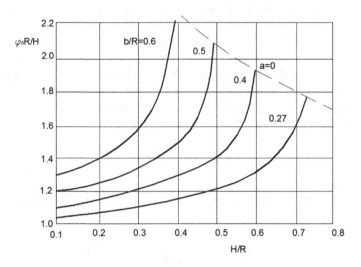

Fig. 3.11 Length of the relative cutting arc for different machine table positions

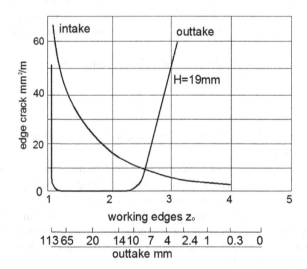

Fig. 3.12 The effect of the tool overhang on the edge breaking for particle board 19 mm thick

3.5 Milling and Planing Machines

The kinematic relations of the milling and planing machines are demonstrated in Fig. 3.13. The tool edge describes a cycloid path in relation to the material similar to circular saws.

The value of the feed per tooth is:

$$e_z = \frac{e}{n \cdot z}$$

The maximum chip thickness is approximately:

$$h_{max} = e_z \cdot \sin \varphi$$

The cutting rotation angle φ—as shown in Fig. 3.14—depends on the H/R ratio. The expression for the φ function is given by:

$$\varphi = 1{,}425\sqrt{\frac{H}{R}}$$

so the average chip thickness:

$$e_{zav} = \frac{e_z \cdot H}{\varphi \cdot R} = \frac{e_z}{1425}\sqrt{\frac{H}{R}} \tag{3.10}$$

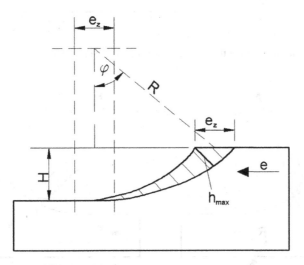

Fig. 3.13 The kinematic relations of a milling machine (Sitkei et al. 1990)

Cutting can be climb cutting or counter cutting in relation to the direction of the feed (Fig. 3.15). With climb cutting the horizontal component of the cutting force is in the same direction as the feed. Therefore, the material can be forced under the tool if the restraining force of the feed is not sufficient. For this reason, it cannot be used for a manual feed. With counter cutting, the horizontal component of the cutting force is opposite to the direction of the feed.

Climb cutting generally produces a smoother surface, because the edge of the knife always exerts compressive stress on the material. With counter cutting, the edge exerts tensile stress on the material that can cause grain laceration fibre. But the great cutting speed of counter cutting due to the inertia forces generally produces the necessary surface quality.

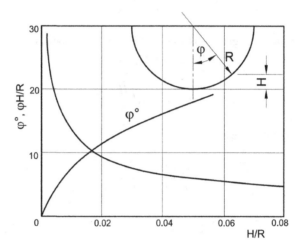

Fig. 3.14 The relative length of the cutting arc depends on the cutting depth

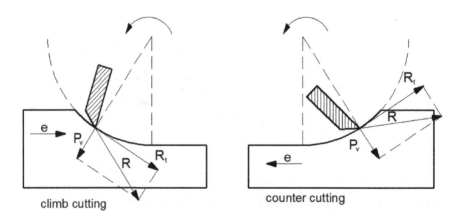

Fig. 3.15 Climb and counter cutting

The inherent disadvantage of the milling and planing operation is the wavy surface caused by the cycloid path (Fig. 3.16). Factors that determine waviness are the wave depth (t_1) which is determined by the diameter of the blade, number of teeth and rotation speed, and the feed speed of the material.

According Pythagoras' theorem, the wave depth can be calculated as follows:

$$(e_z/2)^2 = (D/2)^2 - (D/2 - t_1)^2$$

from which

$$e_z = 2\sqrt{t_1(D - t_1)} \cong 2\sqrt{Dt_1};$$

and

$$t_1 = \frac{e_z^2}{4D} = \frac{e^2}{4Dn^2z^2};$$

From the above equation it is clearly seen that the main governing factor is the tooth bite e_z but also the tool diameter influences the wave depth (geometric roughness). Using cylindrical tools, the radius of tool and that of its path way are the same. As we will just see in the following, there are possibilities to increase the apparent radius of cutting circle of a tool and decrease the wave depth considerably. Wave depth values

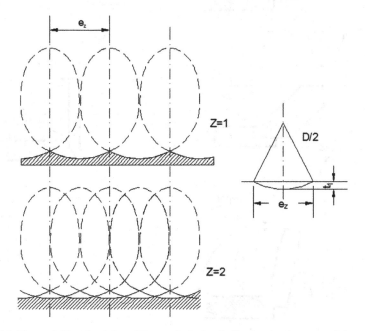

Fig. 3.16 The cycloid path of the milling and planing knife, the development of the waviness (Sitkei et al. 1990)

in the range of 0.1 and 0.5 μm mean a high quality surface while the 0.5–1.2 μm wave depth is considered as a medium quality surface.

In the last decades, several face milling methods were analysed and a new face milling and moulding technique was designed and developed.

The different moulding systems for edge working are classified and shown in Fig. 3.17. The classical method is the use of the moulding cutter head with peripheral blades. There is a 90° angle between the cutter head axis and the plane of the workpiece. The sawing method (hogging saw) is also often used where the tool axis is parallel with the workpiece plane (plane moulding). These two systems may be considered as extreme concerning the setting angle κ. Between these two extreme values, different angles can be used as shown in Fig. 3.17 (Krazhev 1963; Lang 1989).

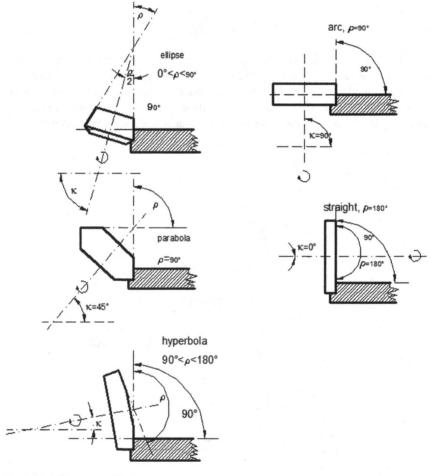

Fig. 3.17 Different moulding systems

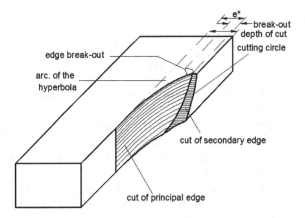

Fig. 3.18 The interaction of the workpiece and the knife

The cutting course varies in different ways depending on the combination of angles which may be ellipsoidal, parabolic or hyperbolic. This last method using a hyperbolic path is very promising to obtain good surface and edge quality and a long service life.

The interaction of the workpiece and knife is shown in Fig. 3.18. (Lang 1989; Tröger and Lang 1990) The path of the principal cutting edge (face cutting edge) is a hyperbolic curve producing a very thin chip at the exit from the wood. The secondary cutting edge (peripheral cutting edge) will cut a part of the maximum depth of the cut. There may be more splintering of the edge at the entry of the knife into the wood, but this break-out will be eliminated by the next cutting path.

The hyperbolic path of the cutting edge may have an enormous cutting circle radius larger than the tool depending on the tool setting angle κ.

Using the notations given in Fig. 3.19, the cutting radius can be calculated by following equation

$$R_k = L \cdot \tan \alpha$$

The distance L is given by

$$L = \frac{D_a}{2 \cdot \sin \alpha} = \frac{R_0}{\sin \alpha}$$

where $D_a = 2R_0$ is the mean diameter of the tool, κ is the setting angle of tool, generally around 5°.

Due to the right-angled triangle, for the sum of the angles α and κ holds

$$\alpha + \kappa = 90°$$

Combining the above two equations yields

$$\frac{R_k}{R_0} = \frac{1}{\cos \alpha} = \frac{1}{\sin \kappa} \tag{3.11}$$

which is plotted in Fig. 3.20. The following example demonstrate the effectiveness of conical tools using large tooth bite.

Taking $R_0 = 75$ mm, $\alpha = 85°$ ($\kappa = 5°$), $H = 2$ mm and $e_z = 4$ mm. The distance L will be 75.3 mm, $R_k = 857$ mm, the geometric roughness $t_1 = 2.3$ μm and the average chip thickness, using Eq. (3.10), is $h_{av} = 0.136$ mm. If we use cylindrical tool of the same diameter, then $t_1 = 26.7$ μm and $h_{av} = 0.4$ mm. The arc length of cutting, however, changes in opposite direction, 59 mm and 13.3 mm respectively. The energy consumption, therefore, more or less remains unchanged.

The cutter head has a setting angle κ and a tilting angle λ in the forward direction, as shown in Fig. 3.21. The tilting angle λ results in the above mentioned hyperbolic path of the knife and formation of thin chips at the exit from the wood. This is essential for a good surface and edge quality. It is also important that the cutting circle will be much larger than the diameter of the cutter head. The larger cutting circle provides more compression forces, which may contribute to less break-out from the edge of the workpiece. The optimum choice of the setting and tilting angles can only be established by experiments. The experimental results are shown in Fig. 3.22 where the edge splintering has minimum values between 5 and 10 degrees of the setting angle. In these experiments the cutter head operated with 1 knife and had a diameter of 200 mm. Furthermore, it is remarkable that the knife can work with a tooth feed of 5 mm with almost no edge break-out.

Having the optimum setting angle range, the next question is to examine the influence of the tilting angle on the quality. The influence of tilting (machine) angle on the edge splintering is shown in Fig. 3.23. Quite a small angle may be selected

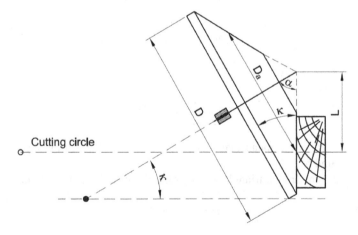

Fig. 3.19 Working principle of a cone-shaped milling cutter

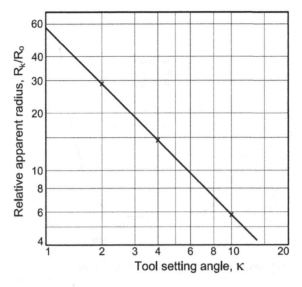

Fig. 3.20 Relationship between the apparent cutting radius and the tool setting angle

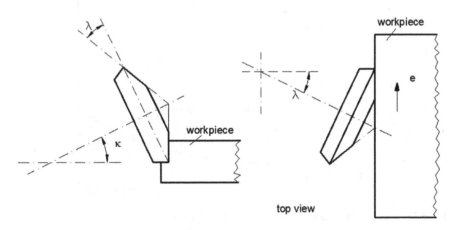

Fig. 3.21 The setting and tilting angles

and the optimum values are from 0.5° to 1.5°. A negative value of λ rapidly increases the splintering of the edge.

Further experiments have shown that the rake angle of the knife may be as low as 5–10°; therefore, TC reversible knives may also be used which makes the system generally applicable. This reduces the wear of the knives. A cutter head with 12 knives equipped with fine grained TC reversible plates has machined 36,000 m long edge surface without needing re-sharpening.

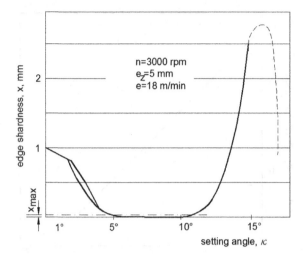

Fig. 3.22 Variation of edge splintering as a function of setting angle

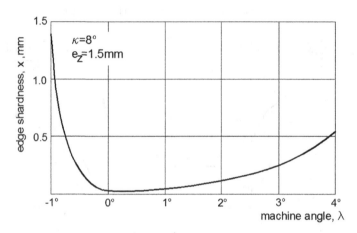

Fig. 3.23 Edge quality variation with tilting (machine) angles

3.6 Edge Machining

Edge machining is a common and important operation in the furniture industry. As
mentioned earlier, edge machining may cause edge breaking which considerably
lowers the quality and aesthetic value of the workpiece. Edge breaking is measured
by the summation of broken surfaces along the unit length in mm^2/m (see in Sect. 8.2).
Edge breaking can be reduced by using a larger tool diameter and thin chips. Especially cone-shaped tools are effective to avoid edge breaking as already outlined
above.

Using the common milling or moulding cutter, the main influencing factors are the feed per tooth e_z and the sharpness of cutting edge or feed distance L_f, Fig. 3.24 (Licher 1991).

The feed distance is calculated as

$$L_f = e \cdot T = e_z n \cdot z \cdot T$$

while the true cutting distance is given by

$$L_c = L_f \frac{R \cdot \varphi}{e_z} \text{ with } R \cdot \varphi = 1.42 \cdot \sqrt{H \cdot R} \tag{3.12}$$

where

T is the tool lifetime,
R is the tool radius,
φ is the angle of cutting,
H is the depth of cut.

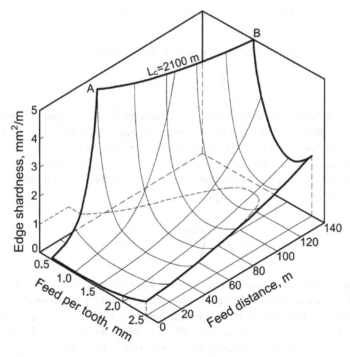

Fig. 3.24 Edge shardness as a function of feed per tooth and feed distance. $D = 180$ mm, $H = 2$ mm, $z = 1$, $v_c = 60$ m/s. Particle board and hard metal edge

It is obvious that the wear of the edge is directly influenced by the cutting distance and not by the feed distance. Evaluation of Fig. 3.24 revealed that along the AB-line $L_c = 2100$ m = constant, which means, constant shardness corresponds to a given constant cutting distance. This recognition facilitates the correct description of the experimental results. From Eq. (3.12) we get

$$\frac{L_f}{e_z} = \frac{L_c}{R\varphi} = 110.230$$

which is constant along the AB line. The edge sharpness due to the wear is governed by the cutting distance L_c given in Eq. (3.12).

The smaller component of shardness as a function of tooth bite e_z, for $L_f = 0$, may be approximated in the following form:

$$S(1) = A \cdot e_z^2$$

while the major component depending on the cutting distance has an exponential character

$$S(2) = e^m - 1 \text{ with } m = K\left(\sqrt{R \cdot H}\frac{L_f}{e_z}\right)^n$$

The resultant edge shardness is the sum of the above two equations

$$S = A \cdot e_z^2 + e^m - 1 \text{ mm}^2/\text{m} \tag{3.13}$$

where A, K and n are constants and they can be determined from experimental results. In Fig. 3.24, the constants have the following values: $A = 8 \times 10^4$, $K = 2.755 \times 10^{-4}$ and the exponent n = 1.2, if metric dimensions are used.

In practical cases the allowable shardness is prescribed and the feed distance L_f is the dependent variable as a function of tooth bite e_z, the cutting depth H and the tool radius R. From Eq. (3.13) we get

$$L_f = \frac{e_z}{\sqrt{R \cdot H}}\left(\frac{\ln\left(S + 1 - A \cdot e_z^2\right)}{K}\right)^{1/n} \tag{3.13a}$$

The above general regularities are suitable to select proper tool and operational parameters to achieve good edge quality and the maximum possible production rate. Further discussion of this subject is given in Sect. 9.4.3.

Sadly, Fig. 3.24 is not generally valid for other wood materials. The form Eq. (3.13), may be used to process arbitrary experimental results.

3.7 Drills

Drills are used to make through-holes and dead-end holes mainly in the furniture industry. The main types of drills are seen in Fig. 3.25a–d. Double-edge twist drills are used to drill through holes. They have a diameter d and a cone angle 2κ. The feed per revolution is given by

$$e_n = \frac{e}{n}$$

where e is the feed rate, m/min, and n is the rotation speed, rpm. The feed per tooth will be calculated as

$$e_z = \frac{e}{z \cdot n}$$

where z is the number of cutting edges.
    The length of the cutting edge is

$$b = \frac{d}{2 \sin \kappa}$$

and the chip thickness is

$$h = e_z \cdot \sin \kappa$$

Drills have a rotation and a feed motion, so their moving clearance angle decreases the true clearance angle (see Fig. 1.3). In the axis of a drill the moving clearance angle is 90°, so the central part of the drill always compresses the material. To reduce the compressed area as much as possible, the ratio of the tangential speed and feed rate (v/e) should be kept between 30 and 60 (Fig. 3.25).

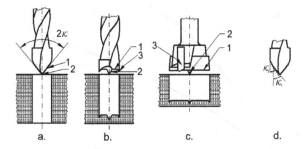

Fig. 3.25 Different machine bits using for through and dead-end drilling **a–d**

This ratio is calculated from the following equation

$$\frac{v}{e} = \pi \frac{d}{e_n} \text{ and } \tan \alpha' = \frac{e}{v}$$

where α' is the moving clearance angle.

The rotation speed of drilling is generally 4000–9000 rpm but higher rotation speeds up to 20,000 rpm can be used.

Higher rotation speeds create some problems drilling dead-end holes, because the deceleration of the feed is limited to 3–4 m/s^2 (Westkämper and Kisselbach 1995).

The surface roughness of the hole depends on the feed per revolution. Drill types **b** and **c** are less sensitive to higher feeds per revolution because they have spur edges. These edges generally have a negative rake angle ensuring a better surface quality.

Special attention should be paid to the quality of the edge of the hole. Boring through holes, edge breaks and deformed edges (burr formation) often occur. The bottom edge is especially sensitive to edge breaking. In many cases, the only method to avoid burr formation is to use scrap wood under the worked material when drilling through holes. This is especially true for particleboard, MDF and covered coated boards.

The edge quality depends on the sharpness of the drill and also on the cone angle. A smaller cone angle gives a better quality edge of the hole. To do this, the cutting edge of the twist drill can be designed with two different cone angles, as shown in Fig. 3.26d. The second major influencing factor is the feed per revolution. A higher feed per revolution may be used for hard woods (beech) than for softwoods.

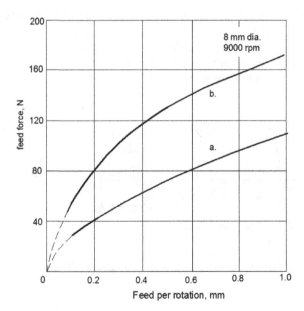

Fig. 3.26 Feed force versus feed per rotation (e) for drill types **a** and **b**. 1—cutting edge; 2—centre drill; 3—spur edge (Westkämper and Kisselbach 1995)

The central part of the drill, generally within the 0.1 R circle area, compresses the worked material. This is associated with a relatively high feed force (Fig. 3.26e). As a rough estimate, the feed force has the same magnitude as the cutting force reduced to the outer radius of the drill.

Due to the small size of drilling tools, the energy requirement of drilling is generally low. Measuring the energy consumption it seems to be advisable to measure the driving torque and not the cutting force.

The power consumption is:

$$P = \frac{M \cdot n}{9.55} = M \cdot \omega$$

where n means the rotation speed, rpm and ω is the angular speed, rad/s. The driving torque can be calculated using Eq. (1.8a) multiplied by the drill radius (the cutting force is related to the outer radius of the tool) in the following way:

$$M = (A + B \cdot h) \cdot b \cdot r = (A + B \cdot h)\frac{r^2 \cdot z}{\sin \kappa} \qquad (3.14)$$

with $b = z\frac{r}{\sin \kappa}$ $h = e_z \sin \kappa$.

Where the angle κ is the half of the drill cone angle, z is the number of edges. Using twist drills with cone angles between 60 and 80°, the constants in the equation have average values of $A = 34$ 1/cm and $B = 1250$ 1/cm^2. In this case the torque is obtained in N cm.

Drills with spur edges (used to produce blind holes) generally have lower torque requirement than twist drills. Using high speed, however, the power requirement may be around 1 kW (diameter 8 mm and $n = 10,000$ rpm).

The cutting energy required for drilling related to the unit volume of solid wood is in the same magnitude as at the common woodworking operations: 17–25 kWh/m^3 (see Table 4.1).

3.8 Veneer Cutting and Peeling

The two main procedures for veneer production are linear and rotary cutting (peeling). The difference between the two cutting processes can be seen in Fig. 3.27. Both use a small bevel angle around 20°; the clearance angle is as small as possible, generally between 0.5 and 1.5°.

The pressure bar is important in the production of quality veneer (see in Sect. 1.6). To produce good quality veneer, the nosebar position, both vertically and horizontally, must be optimally set over the knife. The setting parameters are the *relative deformation* of the veneer (z_0/h) and the *horizontal gap* (lead), which is the distance c between the rear edge of the nosebar and the cutting edge related to the veneer thickness (c/h).

The relative deformation of the veneer under the nosebar can be achieved by a prescribed setting of the nosebar relative to the cutting edge. The prescribed deformation of the veneer z_0 is 10–20% of the veneer thickness. Another possible method may be to pre-stress the nosebar with a given constant load (e.g. spring load). Practical loads for nosebars are 200–300 N/cm.

Since the actual radial force on the nosebar is not constant, it varies in relation to its average value. That is due to in homogeneities of the wood material and to oscillations of the tool. The force amplitude variation related to the average value is between 15 and 30%.

The variation of nosebar radial forces as a function of the relative deformation of veneer 2 mm thick is shown in Fig. 3.28 (Mote and Marchal 2001). As the relative deformation varies, the contact surface of the nosebar does not remain constant; therefore, the radial force increases more than linearly. The horizontal gap of the nosebar also influences the radial force. With a small horizontal gap, the rake face of the tool hinders the free deformation of the wood so the radial force on the nosebar will slightly be higher.

A generalization of the radial nosebar force was found as a reduced force on the vertical axis $F_r/h^{1.5}$ (Fig. 3.29). For very small deformations, especially for softwoods, some deviation from the straight line is obtained due to the surface roughness causing a smaller modulus of elasticity in the thin surface layer. This effect is smaller for hard woods.

The nosebar slides on the wood surface causing friction forces. These forces can be roughly estimated as true friction forces using friction coefficients between 0.32 and 0.38. The friction forces will be added to the cutting force.

The tangential cutting forces depend on the species of wood and cutting parameters. The main cutting parameters are the veneer thickness, the relative deformation of veneer and the horizontal gap (relative lead). The temperature of the wood and the cutting speed has a smaller influence. The specific cutting force related to the

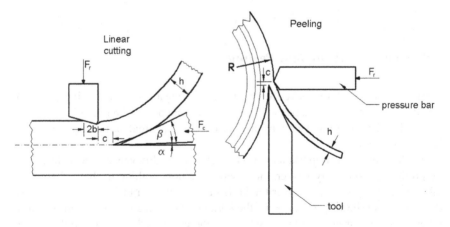

Fig. 3.27 Veneer production with linear cutting and peeling

thickness h (h is given in cm) can be approximated with a straight line as a function of the relative deformation of the nosebar. Measurement results for beech and fir are presented in Fig. 3.30 for veneer with thicknesses between 1 and 3 mm (Mote and Marchal 2001; Bakar and Marchal 2003).

Excessive relative deformation may result in a rough surface and a large power requirement.

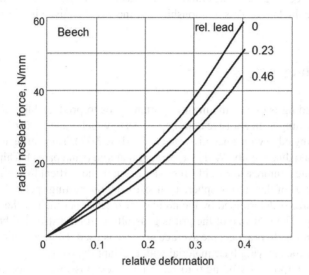

Fig. 3.28 Radial nosebar force depending on relative deformation and relative lead

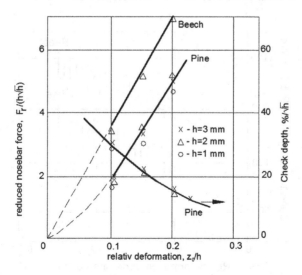

Fig. 3.29 Generalized nosebar forces and check depth as a function of relative deformation (recalculated from Mote and Marchal 2001)

The quality of veneer can be characterised by the surface roughness and the relative checking depth of a veneer surface in tension. The checks depend on the veneer thickness, wood species, temperature of the wood and the nosebar pressure.

The nosebar pressure or relative deformation under the nosebar is a powerful means to reduce surface checking. Check depth is generally represented as a relative value related to the veneer thickness in percent. If this value is divided by the square-root of the veneer thickness, then this reduced check depth is more or less independent of the veneer thickness. This relationship for pine wood is given in Fig. 3.29.

3.9 Sanding

Abrasive sanding is one of the most important steps to produce high quality wood products. Sanding is an abrasive cutting process in which material is removed by an unusual cutting edge with a negative rake angle (Fig. 3.31). The cutting tool is a hard sphere with a radius R embedded in a carrying material (sand paper). Chip formation depends on the compression yield stress of the wood, the vertical load on the cutting edge, the radius of the cutting sphere (grit size) and the cutting speed.

The grit size is determined by the number of meshes per inch of the sieves used for screening. The real size of the grit is generally normal distributed between two successive sieves. In practice, the average grit size is used. Huge amounts of grit are embedded in the carrying base material with overlapping paths.

The main purpose of sanding is to achieve the best possible surface quality characterized by the different *surface roughness parameters* (see in Sect. 8.3). The most important operational parameters are the surface pressure, the feed speed, the grit

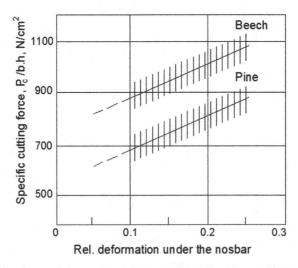

Fig. 3.30 Specific cutting forces for peeling pine and beech, cutting thickness between 1 and 3 mm

size and the sanding speed. The previous machining also has a considerable influence on the operational parameters: the thickness of the layer to be removed strongly depends on the previous machining quality.

The abrasive material (grit) is generally made of either aluminium oxide (Al_2O_3) or silicon carbide (SiC). They have different physical properties in their hardness and heat conduction coefficient. Silicon carbide has a hardness of 9.7 (Mohs) and $\lambda = 41$ W/m K, while aluminium oxide has 9.0 and $\lambda = 20$ W/m K. The high heat conduction coefficient is very important for sanding wood composites containing adhesives. In practice, corundum is mainly used for sanding solid woods but, silicon carbide gives better results on particle board and MDF. At high temperatures, the adhesive may soften causing smearing and clogging. Choosing an abrasive with a better heat conduction coefficient may prevent such difficulties (Dobrindt 1991).

Sanding is a special cutting process characterized by a negative rake angle of the cutting edge and by the random position of grit embedded in the holding tissue. An accurate description of the sanding process is not possible; nevertheless, some general relationships can be derived. The deformation of a spherical rigid indenter acting on a plane surface, is given by the following equation (Timoshenko and Goodier 1951):

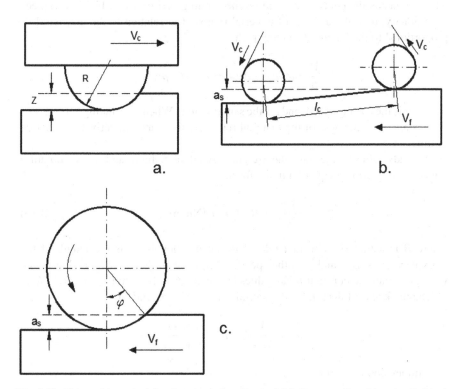

Fig. 3.31 The working principle of a spherical cutting tool (**a**), linear sanding (**b**) and cylindrical sanding (**c**)

$$z = const \cdot \left(\frac{F}{E}\right)^{\frac{2}{3}} \cdot \frac{1}{R^{\frac{1}{3}}}$$

where F is the force, R is the radius of the indenter and E is the modulus of elasticity.

Moving the spherical indenter horizontally, the chip cross section is an arc of a circle with a height z and its value can be approximated with the following simple equation:

$$A = \frac{\pi}{2} R^{0.55} \cdot z^{1.45}$$

Using these two equations, and assuming that grit always has the same relative cover of the surface, the chip cross section assumes the following form (Scholz and Ratnasingam 2005):

$$A \cong const \cdot \frac{F}{E} = const \cdot p$$

where p means the contact pressure on the sanding belt or drum. If the workpiece is stationary, then the amount of material removed in relation to the unit width is proportional to the length of sanding path

$$\frac{V}{b} = const \cdot A v_c t = const \cdot p v_c t$$

where v_c is the cutting speed, and t is the sanding time. When the sanding surface area is constant, the thickness of the material removed also varies linearly as a function of time.

It is advisable to measure the weight instead of volume and so we obtain a dimensionally correct equation in the form:

$$\frac{V \gamma}{b} = B \cdot L p v_c t \ [\text{N/cm}] \tag{3.15}$$

where B is a material constant related to the unit length, L is the length of the workpiece ($L \leq L_c$), and γ is the specific weight of the wood, N/cm^3. For beech wood, the material constant B has values between 4.6×10^{-8} and 5.4×10^{-8} 1/cm.

The thickness of the removed material is also time dependent:

$$a_s = \frac{V}{b \cdot L} = B \cdot \frac{p \cdot v_c \cdot t}{\gamma}$$

or in dimensionless form

$$\frac{a_s}{v_c \cdot t} = B \frac{p}{\gamma}$$

However, the workpiece generally moves under the sanding belt or drum with a feed speed v_f. The sanding time for a given point depends on the contact length L_c and feed speed

$$t = \frac{L_c}{v_f}$$

where L_c is the maximum contact length of the sanding device (see Fig. 3.31). The removed material is now calculated as follows:

$$\frac{V \cdot \gamma}{b} = \frac{B \cdot L_c^2 \cdot p \cdot v_c}{v_f} \tag{3.15a}$$

The thickness of the material removed is calculated as

$$a_s = \frac{V}{b \cdot L_c} = B \cdot L_c \frac{p \cdot v_c}{\gamma \cdot v_f} \tag{3.15b}$$

or in dimensionless form

$$\frac{a_s}{L_c} = B \frac{p \cdot v_c}{\gamma \cdot v_f}$$

The amount of wood removed related to the unit area and unit time is given by

$$k = \rho \frac{a_s}{L_c} v_f \ [\text{g/cm}^2\text{min}] \tag{3.16}$$

where ρ means the density of the wood.

Approximate values of the amount of wood removed for a surface pressure of 1 N/cm^2 is given in Table 3.1.

The amount of wood removed by sanding belts continuously decreases as a function of working time due to the wear process. Wood species containing resins work less effectively in the first 20–30 min.

Table 3.1 Values of k for different wood species

Wood species	Beech	Alder	Oak	Pine
k, g/cm^{2}min	0.25–0.3	0.4–0.5	0.2–0.25	0.3–0.4

Note The above figures are valid for $v_c = 8$–9 m/s

Keeping in mind Eq. (3.15b), the following dimensionless number can be derived:

$$\Pi = \frac{p \cdot v_c}{a_s \cdot \gamma \cdot v_f}$$

which should be invariant for sanding a given wood material. Combining Eqs. (3.15b) and (3.16), the amount of wood removal k can be related to the material constant B as

$$k = B \cdot \rho \cdot \frac{p}{\gamma} \cdot v_c \qquad (3.16a)$$

or

$$B = \frac{k \cdot \gamma}{\rho \cdot p \cdot v_c} \qquad (3.16b)$$

or using Eq. (3.16), we get

$$B = a_s \frac{\gamma}{p \cdot L_c} \frac{v_f}{v_c} \qquad (3.16c)$$

In the possession of some experimental results, e.g. thickness of the removed material a_s, all important operational parameters can easily be calculated for arbitrary initial conditions. The continuous wear process should be taken into account resulting in a decrease of all performance parameters as a function of working time. The simplest way to account for the wear process is to use an exponential function in the form:

$$B = B_0 \cdot e^{-\beta \cdot t}$$

or

$$k = k_0 e^{-\beta t}$$

where B_0 is the initial value of the material constant B and the constant β characterizes the rate of decrease as a function of time. Here the wood removal is time dependent and its value is obtained by integration over the required time span t:

$$\frac{V}{b} = B_0 \cdot L_c \cdot v_c \frac{p}{\gamma} \int_0^t e^{-\beta \cdot t} dt$$

which yields

$$\frac{V}{b} = B_0 \cdot L_c \cdot v_c \cdot \frac{p}{\gamma} \frac{1}{\beta} \left(1 - e^{-\beta \cdot t}\right) \qquad (3.15c)$$

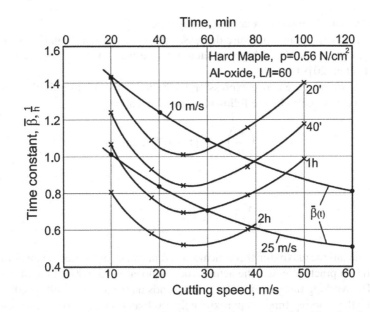

Fig. 3.32 Time constant for sanding as a function of cutting speed and working time (calculated from Hinken 1954)

The value of β is dependent on the belt speed and platen pressure. Sadly, very few good organized experiments have been done and the most extensive research was carried out 65 years ago (Hinken 1954). The new processing of these experimental results is seen in Fig. 3.32 for hard maple with platen pressure of 0.56 N/cm^2. The ratio of belt length to the sanding contact length was 60 which characterises the true contact time. The contact time is defined as that time during which a given area of a belt is in contact with the workpiece. Between the elapsed time t and the true contact time t_r the following simple relationship is valid

$$t_r = \frac{l}{L} \cdot t$$

where l is the length of workpiece sanded and L is the length of the belt. Comparison of different experimental results is only possible taken the true contact time into consideration.

In Fig. 3.32 it is clearly seen that the time constant β has a definite minimum at 25 m/s cutting speed. Furthermore, the time constant considerably decreases during the working time, therefore, the rate of decrease in stock removal only in individual time intervals may be regarded as constant. An equivalent average time constant can be calculated using the particular weighted values.

$$\beta_{av}(t_0) = \sum_{i=1}^{n} \beta_i \frac{t_i}{t_0} \tag{3.17}$$

where t_o means the total (or related) sanding time.

The increase of platen pressure decreases the time constant β and, therefore, a moderate cutting speed and higher platen pressure may ensure a better efficiency (Csanády et al. 2019).

Keeping in mind the above expression for β, the material removal related to the unit width can be given by the following general equation:

$$\frac{V}{b} = \left(\frac{V}{b}\right)_{max} \cdot \left(1 - e^{-\beta \cdot t}\right) \ cm^3/cm \qquad (3.15d)$$

with

$$\left(\frac{V}{b}\right)_{max} = \frac{B_0 \cdot p \cdot v_c \cdot L_c}{\gamma \cdot \overline{\beta}(t_0)}$$

$(V/b)_{max}$ means the maximum theoretical stock removal using the belt for an infinitely long time. In practical cases, the actual stock removal is 75–80% of the maximum value. The working time required for this depends on the sanding belt speed: for $v_c = 20$ m/s the working time is approximately 1.0 h and for $v_c = 10$ m/s the time is doubled.

Exponent β probably depends on the contact pressure. Experiments are needed to clarify this relationship in detail.

Limited experimental results show that the theoretically obtained linearity between stock removal and surface pressure in a wider range does not hold. When the surface pressure increases, more chips are in the contact surface and they may hinder the full penetration of grit into the wood surface corresponding to the given pressure. Therefore, slightly less wood is removed at higher pressures than expected from the linear law. An appropriate correction can be made by modifying the maximum stock removal using an exponent for p less than one:

$$\left(\frac{V}{b}\right)_{max} = \frac{B_0}{\overline{\beta}} L_c \frac{p^{0.8} v_c}{\gamma}$$

Then the material constant B_0 has to be modified and it is between 3.5×10^{-8} and 4.1×10^{-8} 1/cm (beech wood).

The same thing must be done to calculate the effect of grit size on stock removal The free height of fine grit is very low and, therefore, the chips on the sliding surface hinder the penetration of their edges into the wood surface. Experiments show that the wood removal, using P 180 or finer grits, decreases considerably (Dobrindt 1991).

Figure 3.33 shows the wood removal as a function of surface pressure for different grit sizes. The relationship is theoretically linear; in practice, a slight deviation is obtained in the higher surface pressure range.

Wood removal is inversely proportional to the feed speed due to the decreasing time in action on a given surface. Therefore, a strong correlation exists between thickness of the layer to be removed and the feed speed.

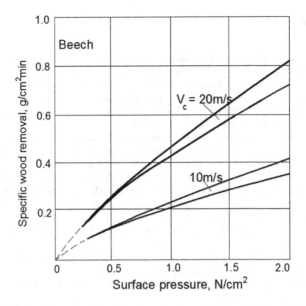

Fig. 3.33 Wood removal as a function of surface pressure. Grit size is P80 and P120

The sanding speed and the grit size of the belt are the most important operational parameters to obtain a good surface roughness. Figure 3.34 demonstrates the effect of grit size on the surface roughness parameter R_z for beech wood. The relationship is practically linear in the given grit size range (Siklinka and Ockajova 2001).

An increased sanding speed reduces roughness, but it is less effective compared to the grit size. An increased sanding speed from 10 to 25 m/s improves the roughness values by $\Delta R_z = 3$ to 4 μm (for more details see in Sect. 8.6).

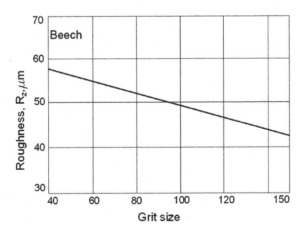

Fig. 3.34 Effect of grit size on the roughness parameter R_z for beech wood. Sanding speed 20 m/s

Literature

Bakar, E., Marchal, R.: Efficiency of automatic adjustment of a slanted-elastic pressure nosebar in peeling. In: Proceedings of 16th IWMS Matsue, pp. 815–821 (2003)

Csanády, E., Kovács, Z., Magoss, E., Ratnasingam, J.: Optimum Design and Manufacture of Wood Products. Springer, Heidelberg, New York, Dordrecht, London, p. 421 (2019)

Dobrindt, P.: Optimiertes Schleifen von MDF- und Spanplatten. Workshop Tagungsband, 8. Holztechn. Koll. Braunschweig, S. 125–136 (1991)

Hinken, E.: Machining wood with coated abrasives. Progress Report, University of Michigan (1954)

Кряжев, Н.А.: Цилиндрическое и коническое Фрезерование древесины (Cylindric and cone-shape milling tools for wood machining). Лесная пром., Москва (1963)

Lang, M.: Untersuchungen zur Entwicklung eines kombinierten Plan- und Umfangsfräsverfahrens mit kegelstumpfförmigen Werkzeugen. Ph.D. Dissertation, Dresden-Sopron (1989)

Lang, M.: A special face milling method for better surface quality and service life. In: Proceedings of 11th Wood Machining Seminar, Oslo (1993)

Licher, E.: Schnittwert-Datenbank für die Holzbearbeitung. 8. Holztechn. Koll. Braunschweig, S. 33–58 (1991)

Mote, F., Marchal, R.: Influence of nosebar settings on tool instabilities in the peeling process. In: Proceedings of 15th IWMS Los Angeles, pp. 309–328 (2001)

Scholz, F., Ratnasingam, J.: Optimization of sanding process. In: Proceedings of 17th IWMS Rosenheim, pp. 422–429 (2005)

Siklinka, M., Ockajova, A.: The study of selected parameters in wood sanding. In: Proceedings of 15th International Wood Machining Seminar Los Angeles, pp. 485–490 (2001)

Sitkei, G., Horváth, M.: Log dynamics during sawing in frame saws. In: Proceedings of 12th IWMS Kyoto, pp. 327–334 (1995)

Sitkei, G., et al.: Theorie des Spanens von Holz. Fortschrittbericht No. 1. Acta Fac. Ligniensis Sopron (1990)

Timoshenko, S., Goodier, J.: Theory of Elasticity. McGraw-Hill Education, New York (1951)

Tröger, J., Lang, M.: Hobelfräsen mit vermindertem Wellenschlag. HOB 11, S. 43–49 (1990)

Westkämper, E., Freytag, J.: PKD-Schneidstoff zum Sägen melaminharzbeschichteter Spanplatten. IDR 1, S. 46–49 (1991)

Westkämper, E., Kisselbach, A.: Leistungsteigerung beim Bohren. HOB, S. 94–103 (1995)

Chapter 4
The Energy Requirement of Cutting

4.1 Introduction

The energy consumption of woodworking operations is significant part of the total production cost. Therefore, it is important to select proper tools with optimum operational parameters.

After a short theoretical consideration, the energy requirements of different woodworking machines are treated. Generally, working diagrams are constructed to facilitate the consideration of several variables influencing the specific force or energy requirements. Generally valid relationships are also included.

In many cases other important features such as production rate, surface quality and tool maintenance cycle must be considered to minimize manufacturing costs. All operating parameters must be analysed to determine the best specifications for each production line.

4.2 Theoretical Considerations

In the discussion of the theory of cutting we derived and proved that the cutting force per unit length can be expressed with the simple linear equation.

$$\frac{P_h}{b} = A + B \cdot h$$

The A and B constants in the equation depend on the mechanical properties of wood, the direction of cut and the cutting angle δ (or rake angle γ) of tool. Figure 4.1. shows the variation of cutting forces for pine and beech woods as a function of chip thickness for different directions of cut and cutting angles (Saljé and Dubenkropp 1977).

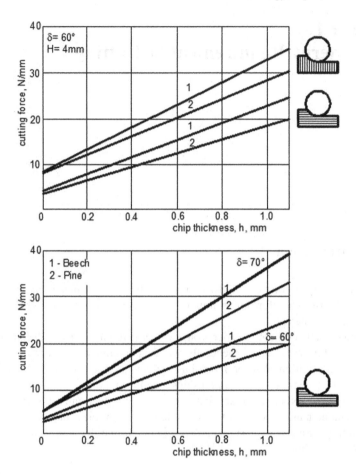

Fig. 4.1 The change of the cutting force depending on the cutting direction and cutting angle δ, for beech (1) and pine (2)

Straight lines are obtained as a function of chip thickness in accordance with the cutting theory. The cutting direction and cutting angle have considerable influence on the cutting force. The increase of cutting angle in a given range, however, increases the surface quality. This effect can be explained by the increase of compressive forces in the surrounding of the edge.

The cutting angle δ (or the rake angle γ) primarily influences the value of the B constant in the cutting equation. As theoretical calculations suggest (see Fig. 1.8.) and experimental results prove, the cutting force strongly depends on the cutting angle, Fig. 4.2 (calculated and replotted from Koch 1964). The constant B can be given by the following empirical equation.

$$B = B_0 + a(\delta - 40)^n$$

where

B_0 is the minimum value of constant B at $\delta = 40°$.

Processing the experimental data gave the following values to the constants:

for beech $a = 1.5*10^{-3}$ N/mm^2 and $n = 2.75$;
for pine $a = 5.5*10^{-4}$ N/mm^2 and $n = 3$.

Some experimental results have indicated that for small chip thickness ($h = 0.1$–0.2 mm) the optimum cutting angle shifts to slightly higher values falling in the range of 45–55 degree (Sanyev 1980).

Certain tools, especially saws, have some special features. Previously we have not dealt with the role of the secondary cutting edge and we have referred the cutting force per unit length to the length of the principal cutting edge. In the most cases this does not cause error.

The main cutting edge of saws is relatively short and the length of the secondary cutting edge in relation to the length of the main cutting edge (15–30%) is not negligible. The work of the secondary cutting edges of a saw tooth is shown in Fig. 4.3. Shearing occurs on two sides of the tooth at the deformation zone of the cutting (F cross section). The width of the sheared cross section depends on the rupture strain. Its value has not yet been exactly determined. But it is probable that the sheared cross section is proportional with chip thickness h, so the value of the shearing force is:

$$P_h^{''} = 2 \cdot c \cdot h \cdot \tau$$

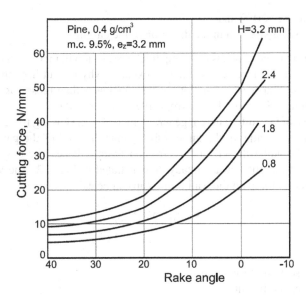

Fig. 4.2 Effect of the rake angle on the specific cutting force at various cutting depths. Tooth bite 3.2 mm

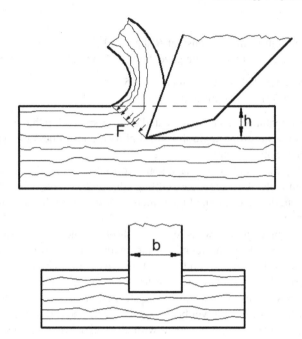

Fig. 4.3 The work of the secondary cutting edge of a saw tooth

where:

c is a constant ($c \cong 0.3$) and
τ means the shear strength in the direction of the cutting.

This shearing force depends on h chip thickness, so it will increase the B constant in the general cutting force equation. This explains why the cutting force diagrams of the saws are steeper as a function of chip thickness than those for milling and planing machines (Salje and Dubenkropp 1977; Sanyev 1980).

An increase of the cutting height generally increases the cutting force. The reason is that the mass of chips in the gullet rubs against the wall of the gap. The more chips are in the gullet the greater the friction force will be. The development of side pressure in the gullet was already demonstrated in Fig. 3.2. The value of the additional friction force on the wall of the gap can be calculated as

$$S = \mu \cdot 2 \cdot F \cdot \sigma_h$$

where:

F the cross section of the gullet
μ the friction coefficient
σ_h the side pressure of the chips in the gullet

The volume of the chips can be decreased by 20–30% with relatively small pressure. But a greater compaction in the gullet substantially increases the side pressure and the friction force.

The amount of chips in the gullet can be expressed with the *gullet feed index*:

$$\varphi = \frac{V_{chip}}{V_g} = \varepsilon \frac{V_{sw}}{V_g}$$

where:

V_{sw} volume of solid wood
V_{chip} volume of chips

The φ gullet feed index has the unit value if loose chips fill the gullet. If the chips are totally compacted, then $\varphi = \varepsilon$ ($\varepsilon = 3$–4).

The power requirement can be generally written as follows:

$$N = N_0 + P \cdot v$$

where:

N_0 the idling power
P all of the forces
v the speed

The force is approximately proportional to the chip thickness or to the feed per tooth at cutting, while the number of teeth in action is:

$$z' = \frac{H_\Sigma}{t}$$

where:

H_Σ sum of the cutting heights
t the tooth-pitch

Keeping in mind Eq. (3.1), the power requirement can be written as follows:

$$P \cdot v = const \cdot e_z \cdot v = const \cdot e \cdot H_\Sigma$$

And with this the total power requirement is given by the following general equation:

$$N = N_0 + const \cdot e \cdot H_\Sigma \tag{4.1}$$

This equation shows that it is advisable to use the variable $e.H_\Sigma$ (m²/min) in processing experimental results (Sitkei et al. 1988).

4.3   The Energy Requirements of Bandsaw

The bandsaw is a generally used machine in the wood industry for cutting, profile cutting, ripsawing and splitting logs. Equation (3.3), in which most of the variables in a given case are a constant can be written shortly as follows:

$$e_{max} = \frac{const.}{H}$$

This means that the maximum feed speed is determined by the height of the sawn cross section.

Cutting of a given cross section H the gullet feed index increases linearly and it reaches its maximum value at the exit from the wood. In the same situations, the value of φ_{max} is larger as the height of the sawn cross section increases. The cutting resistance of Scotch pine is demonstrated in Fig. 4.4 depending on the feed per tooth at different cutting heights (Sitkei et al. 1990). In the figure the constant lines of φ_{max} are included and as a comparison to the cutting forces, the cutting resistance of a milling machine with a similar cutting angle is also given.

The equation of the curves can be given in the following fashion:

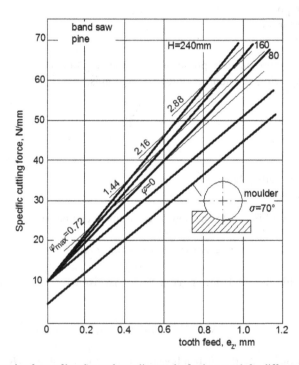

Fig. 4.4   The cutting force of bandsaws depending on the feed per tooth for different cutting heights

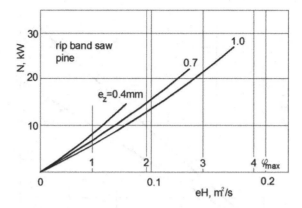

Fig. 4.5 Power consumption of the bandsaw depending on the sawn cross section in the unit time

$$\frac{P_h}{b} = 9 + 43 \cdot e_z + 7 \cdot \sqrt{e_z} \cdot \varphi_{max}^{0.9} \qquad (4.2.)$$

In the above equation, a new term is added to take into consideration the friction forces arising on the side walls.

Figure 4.5 shows the power requirement of cutting depending on the sawn cross section in the unit time. The power requirement is a little bit smaller for a thicker chip. That can be explained by the effect of the first term (A) in the cutting force Eq. (1.8a) which will be smaller with a larger thickness of the chip. The equation of the curves in the Figure can be given as follows:

$$N = 146 \cdot (e \cdot H)^{1.15} + 160 \cdot (e \cdot H)e^{\frac{-e_z}{0.65}} \qquad (4.3)$$

The sharpness of the tool substantially influences its cutting force. The data refers to sharp tools, when the radius of the edge is between $\rho = 20$–$30\,\mu m$; (see Sect. 7.2).

4.4 The Energy Requirements of Frame-Saw

Frame-saws have been the base machine of the primary wood processing for more than 100 years. The frame-saw has some disadvantages in spite of its mature technical design. Its moving masses can be balanced only in the first order. Therefore its speed is only 320–340 rpm today and its average cutting speed is $v_{av} = 6$–$7\,m/s$.

The most important factor is the height of the cutting H. The pitch is t = 25–30 mm; the cross section of the gullet is 2.3–2.8 cm², so the maximum feed speed can be given approximately as follows:

with spring set teeth

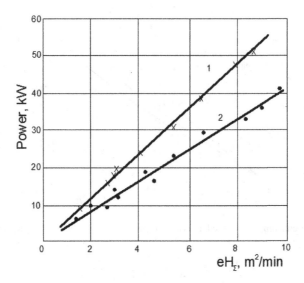

Fig. 4.6 The power consumption of a frame saw depending on the total sawn cross Section 1 black locust, 2 pine

$$e_{max} \cong \frac{0.95}{H} \text{m/min}$$

with swage set teeth

$$e_{max} \cong \frac{0.62}{H} \text{m/min}$$

where the H cut height must be substituted in m.

The idling power of a frame-saw depends on the driving method of the auxiliary equipment. Modern machines are provided with hydraulic devices, and with these we can calculate with $N_0 = 20$ kW value. The cutting power requirement for black locust and Scotch pine are shown in Fig. 4.6. (Sitkei et al. 1988). The power requirement of a frame saw for pine is given by:

$$N = 20 + 4.1 \cdot e \cdot H_\Sigma \text{ kW}$$

for black locust:

$$N = 20 + 6 \cdot e \cdot H_\Sigma \text{ kW}$$

where eH_Σ must be substituted in m²/min.

In order to calculate the total cutting height, the sawing pattern (cutting distance) of the log is needed (Fig. 4.7).

The total cutting height can be calculated as

$$H_\Sigma = 4 \cdot \left[\frac{R}{2} + \sum_{i=0}^{k} \sqrt{R^2 - (a + i \cdot c)^2} \right]$$ (4.4)

where $R \geq a + i \cdot c$ and i is the number of cuts to the right of the centre line.

The theoretical sawing capacity of a frame saw is influenced by the diameter of the log and the feed speed

$$V = \frac{d^2 \pi}{4} \cdot e \cdot 60 \ (\mathrm{m^3/h})$$

Keeping in mind that the feed speed decreases with increasing log diameter, therefore, the sawing capacity is proportional to the log diameter. As a rough estimate, the sawing capacity is given by the following equation

$$V = \eta_t \cdot 45 \cdot d \ (\mathrm{m^3/h})$$

where the log diameter d should be substituted in m, and η_t is the time exploitation coefficient.

A working diagram is needed to facilitate the selection of the optimum operational parameters (Fig. 4.8, Sitkei et al. 1988). The starting point is the log diameter and from this we obtain the maximum feed speed and the cut surface in the unit time. Furthermore, in the lower part of the diagram we find the power requirements for different species of wood.

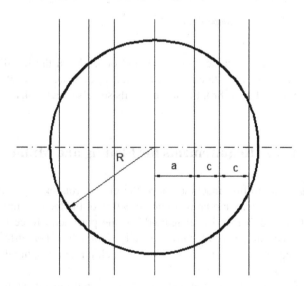

Fig. 4.7 Cutting pattern of frame saws

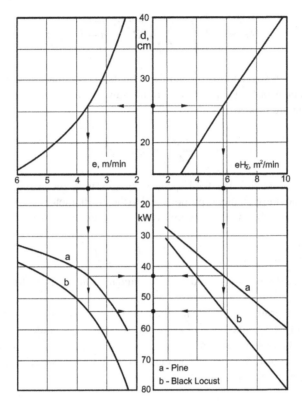

Fig. 4.8   Working diagram for frame-saws

The power to cut one cubic meter of wood depends on the log diameter and increases quickly under $d = 25$ cm. We can usually calculate 4–5 kWh/m^3 specific energy input (Sitkei et al. 1988). For more details see also Sect. 9.4.1.

4.5   The Energy Requirements for Planing and Milling

The planer and the milling machines have the classic rotating tools in the wood industry. We can use them in relation to the direction of the feed in climb or counter cutting run; (see Fig. 3.15). The magnitude of the peripheral force is practically independent of the run, but climb cutting usually results in higher surface quality.

The power requirement depends mainly on the chip thickness, number of edges, the tool radius and the rotation speed.

Using the force shape function and its average force ratio Ψ as introduced in Fig. 2.2b, the power consumption is given by the following equation

$$P = \psi \cdot R \cdot z \cdot b \cdot (A + B \cdot h_{max}) \frac{n}{9.55} \frac{\varphi_0}{360} \tag{4.5}$$

where z means the number of knives and b is the cutting width. The cutting angle φ_0 must be substituted here in degree.

The maximum chip thickness is calculated approximately from the equation

$$h_{max} \cong e_z \cdot \sin \varphi_0 = e_z \cdot \sin \left(1.425 \sqrt{\frac{H}{R}}\right) \cong e_z 1.425 \sqrt{\frac{H}{R}}$$

The average chip thickness is calculated from the chip surface divided by its length:

$$e_{zav} = \frac{e_z \cdot H}{R \cdot \varphi} = \frac{e_z}{1.425} \sqrt{\frac{H}{R}} \tag{4.6}$$

or

$$e_{zav} \cong \frac{h_{max}}{2}$$

The relative value of rotation angle of cutting φ_0 can be expressed as

$$\frac{\varphi_0}{360} = \frac{1.425}{2\pi} \sqrt{\frac{H}{R}} = 0.2268 \sqrt{\frac{H}{R}}$$

indicating that the relative duration of rotation per revolution is a function of the relative depth of the cut.

Example: beech wood is milled with a rotating tool 120 mm in dia., 10 mm cutting width, 4 knives and cutting depth of 4 mm. The maximum chip thickness is $h_{max} = 1$ mm and the rotation speed is 6000 rpm. Using these input data, the cutting angle $\varphi_0 = 0.3679$ rad or $\varphi_0 = 21°$, the tooth bite $e_z = 2.8$ mm and the feed speed is $e = 67.2$ m/min. The cut cross-section in the unit time $eH = 0.27$ m²/min. The calculated power consumption is 1.7 kW. To turn 1 m³ solid wood into chips requires 10.5 kWh/m³ energy for $H = 4$ mm cutting depth, but 19 kWh/m³ for $H = 1$ mm. Using smaller tooth bites, the specific energy consumption further increases.

Figure 4.9 gives a general overview of the energy requirement as a function of operational parameters. Cutterheads with 10–12 knives may have power requirements around 5–6 kW.

Special profile tools used in window frame production have a greater cutting depth and higher cut cross-section in the unit time. Accordingly, they require more driving power compared to the calculated value in the above example.

Figure 4.9 shows clearly that using smaller tooth bite, the power requirement for the same cut cross-section in the time unit increases. Because the effect of cutting depth on the power requirement is not decisive, therefore, for a quick estimate, the

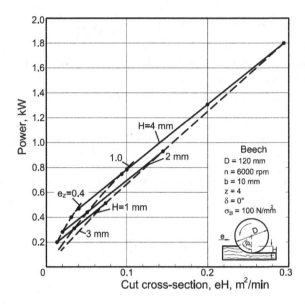

Fig. 4.9 The power consumption of milling

following simple equation may be used for pinewood:

$$P = 0.2 + 5.62(eH) \text{ kW} \tag{4.7}$$

Wood species as a function of their **density** also require different cutting energy. Figure 4.10 shows experimental results for different volume densities as a function of cut cross-section in the unit time. The curves may be described by the following equation:

$$P = 0.7 + 4.75\rho_v^{0.85} \cdot (e \cdot H)^{0.75} \text{ kW} \tag{4.8}$$

where the volume density ρ_v must be substituted in g/cm^3. With smaller cut cross-sections, the tooth bite e_z is generally smaller, which increases the power requirement to a certain extent. The exponent of (e. H). Less than one combines this effect implicitly.

4.6 Energy Requirements of Circular Saws

The working conditions of the circular saw is slightly similar to the band saw, but due to its cycloid movement, the cross section of the chip is variable; (see Fig. 3.10).

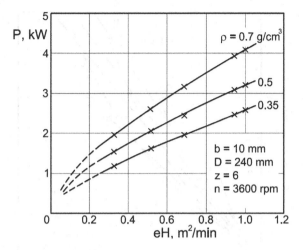

Fig. 4.10 Power requirement for planing as a function of cut cross section in the unit time for different wood densities (recalculated from (Koch 1964))

Forces acting on the main cutting edge are similar to forces acting on milling tools. In addition, forces acting on the secondary cutting edges and chip friction on the gap wall should be taken into account (see in Sect. 4.2).

The resultant tangential force can be calculated as follows:

$$F_t = z'(A \cdot b + B \cdot b \cdot h_{av} + 2c \cdot h_{av}\tau) \cdot \left(\frac{\varphi \cdot R}{H}\right) \qquad (4.9)$$

where:

h_{av} the average chip thickness
z' the number of teeth in action
τ shear strength

For the sake of simplicity, we assumed that the frictional forces increase with the relative arc length of the cutting. The third term in Eq. (4.9) increases the cutting force as a function of chip thicknesses as outlined in Sect. 4.2.

Using Eq. (4.9), the power consumption will be calculated as:

$$P = F_t \cdot \frac{v}{1000} \quad (kW)$$

The power consumption of cutting as a function of the cut cross section is illustrated in Fig. 4.11 for different cutting heights. We have drawn the constant e_z lines on the figure as well. We can see that the energy input (kWh/m^2) increases quickly as the cutting height increases and the feed per tooth decreases.

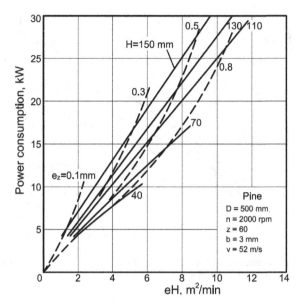

Fig. 4.11 Power consumption of a circular saw. Bending strength of the pine: $\sigma_b = 60$ N/mm^2

The energy consumption of circular sawing also depends on the cutting speed. Figure 4.12 shows experimental results using two different cutting speeds: (fir, cutting height 50 mm, $D = 400$ mm, $s = 2$ mm).

Higher cutting speeds require slightly more power for the same sawing performance due to higher cutting forces and ventilation drags.

The friction forces associated with cutting will be turned into heat, warming up the tooth zone. Due to heat conduction and heat radiation to the air, a typical temperature profile will be developed and its shape can generally be described by a third power

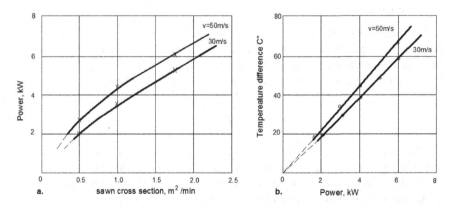

Fig. 4.12 Power requirement as a function of sawn cross section (**a**) and the relationship between the maximum temperature difference and power requirement (**b**) (Sanyev 1980)

Table 4.1 Approximate average values for pine and beech woods, (kWh/m^3 solid wood)

Machines	Pine wood	Beech wood
Bandsaw	20–22	25–28
Frame saw	20–22	25–28
Circular saw	18–20	24–27
Moulder	35–40	44–48
Disc sanding	70–100	80–120
Belt sanding	160–200	200–250

equation. This uneven heat distribution generates compressive tangential stresses in the saw blade reducing the quality of the sawing. The maximum temperature difference along the blade radius is generally proportional to the cutting power as demonstrated in Fig. 4.12b. The cutting speed has some influence on the maximum temperature difference.

Finally, an interesting comparison can be made concerning the energy of chip formation (kWh/m^3 solid wood) for different woodworking machines. Table 4.1 shows approximate average values for pine and beech woods.

Saws work with the same principle and their energy consumption differs very little. Moulding, milling and planing require more energy due to the mixed cutting directions. Sanding has the highest energy requirement, which may be from 5 to 10 times higher in comparison to saws. Especially belt sanders consume much energy due to friction forces on the belt and chip deformation between belt and workpiece.

Solid woods have a strong anisotropy and, therefore, the cutting force significantly differs in the main anatomical directions (see Fig. 4.1). Particle boards and fibre boards are more homogeneus materials requiring no distinction in cutting directions. At the same time, the type and quantity of gluing materials, the density of the boards may cause significant differences in the cutting force and energy requirements (Saljé and Dubenkropp 1983; Salje and Drückhammer 1984; Saljé et al. 1985).

4.7 Energy Requirement of Sanding

Sanding is an abrasive cutting process with very complicated mechanics associated with different friction forces. The accurate calculation of its requirements is almost impossible and, therefore, the following calculation method gives an approximate estimation of the expected energy requirement supporting with experimental results.

The resultant sanding force consists of true cutting forces but the dominant part of these forces originate from various friction components. Therefore we use an equation based on friction forces in the following form:

$$F = (f_e + f)p \cdot A \quad (\text{N}) \tag{4.10}$$

where f_e is an equivalent friction coefficient on the contact surface including also the effect of cutting forces,

f is the friction coefficient on the back side,

p is the surface pressure,

A is the contact surface area.

The equivalent friction coefficient partly depends on the strength of the wood and also on the related thickness of material removal which is characterized with a dimensionless number (see Sect. 3.9):

$$C = \frac{a_s}{v_c \cdot t} = B\frac{p}{\gamma} \quad \frac{cm^3}{cm^2 \cdot cm} \tag{4.11}$$

where a_s is the thickness of removed material,

v_c is the cutting speed,

γ is the volume weight of the wood,

B is a material constant related to the unit width,

t is the elapsed time.

For moving workpieces with a feed speed v_f the above equation has the form

$$C = \frac{a_s}{L_c} = B \cdot \frac{p}{\gamma} \cdot \frac{v_c}{v_f} \quad \frac{cm^3}{cm^2 \cdot cm} \tag{4.11a}$$

Because the equivalent friction coefficient contains true friction, its value is put together from two parts

$$f_e = \frac{2.3 \cdot \sigma_b \cdot C}{p} + f_1 = \frac{2.3 \cdot \sigma_b \cdot B}{\gamma} + f_1 \tag{4.12}$$

where f_1 is the true frictional part of the total resistance coefficient which can be taken as $f_1 = 0.4$ based on experimental results. The first component which substitutes the cutting force is proportional to the bending strength of the wood and the amount of specific material removal. The constant is given from experimental measures (Bershadskiy and Cvetkova 1975).

The power consumption is obtained by multiplying the force with the cutting velocity:

$$P = F \cdot v_c = \left(\frac{2.3 \cdot \sigma_B \cdot B}{\gamma} + f_1 + f \right) \cdot p \cdot b \cdot L_c \cdot v_c \; \text{Watt} \tag{4.13}$$

where the friction coefficient f on the back side has values between 0.3 and 0.4.

Calculations show that the contribution of cutting forces to the total resistance coefficient is relatively small (between 0.15 and 0.3) and the main influencing factors are the friction coefficients, the surface pressure, the contact area and the cutting velocity.

Based on experimental results, the specific wood removal (g/cm^2.min) is often known. Table 3.1 shows average values for stock removal and the corresponding B values for 1 N/cm^2 surface pressure and 10 m/s cutting velocity (see Sect. 3.9).

Example: beech wood is sanded with 1 N/cm^2 surface pressure and 20 m/s cutting velocity. The volume weight is 0.007 N/cm^3, bending strength 10,500 N/cm^2, friction coefficients $f_1 = 0.4$ and $f = 0.35$, material constant $B = 4.8.10^{-8}$ 1/cm. The thickness of the removed material is $a_s = 8.23$ mm/min, the specific stock removal $k = 0.576$ g/cm^2.min. Taking the sanded area as 100 cm wide and 30 cm long, the power consumption is 55 kW which corresponds to 183 kW/m^2. The specific stock removal is 345.6 kg/m^2.h or 1.89 kg/kWh. The removal of 1 m^3 of solid wood requires 370 kWh energy which is at least a magnitude higher than a knife machining.

For comparison, a pine wood is sanded with the same operational parameters, and $\sigma_b = 8000$ N/cm^2, $\gamma = 0.005$ N/cm^3. The thickness of removed materials is $a_s = 12.96$ mm/min and $k = 0.648$ g/cm^2.min, the power requirement is $P = 57$ kW or 189.7 kW/m^2. The specific stock removal is 388.8 kg/m^2.h or 2.05 kg/kWh. The removal of 1 m^3 solid wood consumes 244 kWh which is considerably less than the energy requirement of hardwoods.

Equation (4.13) can easily be rearranged to a dimensionless form. Because the contribution of cutting forces is not decisive, for rough estimation the following simple equation can be used

$$\frac{P}{p \cdot A \cdot v_c} = 0.95 \qquad (4.13a)$$

where $A = b.L_c$ is the sanded surface. The dimensionless number on the left side can be regarded as a similarity number which is more or less invariant for different sanding processes. Using sanding machines with a moving workpiece, Eq. (4.11a) should be used in Eqs. (4.12) and (4.13).

If the thickness of removed material a_s is prescribed, then, using Eq. (4.11) or (4.11a), first the necessary time or surface pressure should be established and thereafter in Eq. (4.13) the value of B/γ should be replaced by

$$\frac{B}{\gamma} = \frac{a_s}{p \cdot v_c \cdot t} \quad \text{or} \quad \frac{B}{\gamma} = \frac{a_s \cdot v_f}{L_c \cdot p \cdot v_c}$$

The performance characteristics of the abrasive belt considerably decreases as a function of working time due to the continuous wear process. The power consumption also decreases but to a lesser extent than the stock removal.

The decrease of stock removal as a function of working time can be taken into account by using time dependent material constants as described in Sect. 3.9.

For a quick estimate, Fig. 4.13 shows approximate specific power requirement as a function of cutting speed and platen pressure.

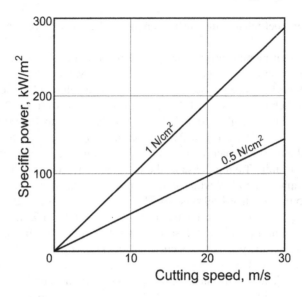

Fig. 4.13 Specific power requirement of sanding

4.8 Generalized Relationships for Energy Requirement

Based on the theorem of Buckingham (Buckingham 1914) and using the standard dimensional analysis method (Langhaar 1951), more generally valid relationships may be established in the form of similarity equation. For planing and milling the following similarity equation has been derived

$$\frac{P}{\sigma_b e b H} = 0.172 \left(\frac{e_z}{R}\right)^{-0.35} \tag{4.14}$$

where σ_b is the bending strength of wood characterizing the cutting resistance (see Sect. 2.2). Equation (4.14) is demonstrated in Fig. 4.14 using calculated and experimentally obtained results.

Although the kinematic relations of different sawing machines (frame, band and circular saws) differ substantially, their cutting mechanism and chip formation are similar. Therefore a successful attempt has been made to derive a similarity relationship to describe the average energy consumption of different sawing machines (Sitkei 2013).

Using the standard dimensional analysis method (Langhaar 1951), the following similarity equation is obtained:

$$\frac{P}{\sigma_B \cdot e \cdot b^2} = const \left(\frac{H}{b}\right)^n \cdot \left(\frac{\varphi_0 \cdot R}{H}\right)^m \tag{4.15}$$

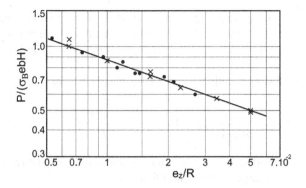

Fig. 4.14 Similarity plot of energy consumption for milling and planing

where

P is the power consumption, Nm/s

σ_B is the bending strength of the wood, N/m^2

e is the feed speed, m/s

b is the width of cutting (kerf), m

H is the sawn height, m

The relative cutting arc. $\varphi_0 R/H$, for saws with a linear motion (frame and band-saws) should be taken as one. The processing of experimental result with different machines and wood resulted in Fig. 4.15 (Sitkei 2013).

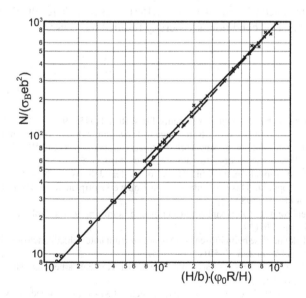

Fig. 4.15 Similarity relationship for frame, band and circular saws

The diagram shows two curves lying very close to each other. The lower part represents the measurement data for circular saws and the upper part for frame and bandsaws. The tangents of curves are 1.08 and 1.04 respectively and, for the sake of simplicity, both may be taken as $n = m = 1.08$.

With this small simplification Eq. (4.15) has the final form:

$$\frac{P}{\sigma_B \cdot e \cdot b^2} = 0.5 \cdot \left(\frac{H}{b} \cdot \frac{\varphi_0 \cdot R}{H} \right)^{1.08} \tag{4.15a}$$

For frame and band saws, for which the relative cutting arc is one, the similarity equation can be rearranged to this simplified form

$$\frac{P}{\sigma_B \cdot e \cdot b \cdot H} = \text{constant} \tag{4.15b}$$

indicating that this dimensionless number is invariant for these sawing machines. For a given saw and wood species, σ_B and b are constant and Eq. (4.15b) yields the simple relationship

$$P = const \cdot e \cdot H \tag{4.15c}$$

which was deduced earlier in another way (Sitkei et al. 1988). In practice, the use of Eq. (4.15c) or a similar plot is more convenient. It is valid for a given tool and a change in the tool with different cutting width requires another constant in Eq. (4.15c).

It should be noted that Eq. (4.15) or (4.15a) gives reliable results if the tooth bite is not too small. This condition is generally fulfilled for saws.

Literature

Buckingham, E.: On physically similar systems. Phys. Rev. **4**, 345 (1914)

Csanády, E., Kovács, Z., Magoss, E., Ratnasingam, J.: Optimum design and manufacture of wood products. Springer, Heidelberg, New York, Dordrecht, London, p. 421 (2019)

Koch, P.: Wood Machining Process. Ronald press, New York (1964)

Langhaar, H.: Dimensional Analysis and Theory of Models. John Wiley, New York (1951)

Saljé, E., Dubenkropp, D.: Das Kantenfräsen von Holzwerkstoffplatten. Holz-und Kunstoffverarbeitung **4**, S. 490–494 (1983)

Saljé, E., Dubenkropp, D.: Zerspanbarkeitskennwerte beim Fräsen von Holz. Holz als Roh-und Werkstoff, S. 333–336 (1977)

Saljé, E. et al.: Erkenntnisse beim Fräsen von Spanplatten mit unterschiedlichen Schnittbedinungen. Holz als Roh-und Werkstoff, S. 501–506 (1985)

Saljé, E. Drückhammer, J.: Qualitätskontrolle bei der Kantenbearbeitung. Holz als Roh-und Werkstoff, S. 187–192 (1984)

Sitkei, G.: Similarity study oft he energy requirement of saws. In: Proceedings Of 21st IWMS Conference Tsukuba, pp. 195–205 (2013)

Sitkei, G. et. al.: Schnittleistung und Energiebedarf von Gattersägen. Acta Fac. Lign. S. 23–31 (1988)

Sitkei, G. et. al.: Theorie des Spanens von Holz. Fortschrittbericht No.1. Acta Fac. Ligniensis Sopron (1990)

Санев, В., Обработка древесины круглими пилами. Изд. Лесная Пром (Woodworking with circular saw) (1980)

Бершадский, А.-Н. Цветкова, Резание древесины. (Machining of wood) Минск (1975)

Chapter 5
Vibration of the Tools and Workpieces

5.1 Introduction

Due to the elastic support and elastic nature of wood, the machine, the tool and the workpiece generally create vibration systems. The natural frequency and the deflection of the free vibration will be determined by the mass and the spring constant of the support. The more rigid the vibration system is, the smaller the deflections will be.

Woodworking machines standing on a foundation form a vibration system. The vertical displacement of the machine table is usually 15–30 μm, and this does not significantly influence the accuracy of the woodworking and the surface quality. An exception is the frame-saw, which has remarkable vibration amplitudes due to the unbalanced masses of the sliding crank mechanism but a frame-saw is seldom used for doing precise work.

It is important to decrease the vibration of saws to achieve good sawing accuracy and surface quality. Saws are thin walled structural elements, and their bending stiffness is small. Therefore, significant lateral and torsional vibrations can develop.

One of the unique features of woodworking machines is that the workpiece moves during processing. The moving wood is pressed to the machine table by spring loaded rollers, and the wood vibrates against them.

In a CNC machine, the workpiece is fixed by pneumatic clamping devices which are not entirely rigid.

In the following, the vibration behaviour of the different woodworking machines will be treated in detail. Calculation methods and experimental results will be presented with practical recommendations.

E. Csanády and E. Magoss, *Mechanics of Wood Machining*, https://doi.org/10.1007/978-3-030-51481-5_5

5.2   Band Saw Vibration

The schematic layout of a band saw and the induced vibration patterns can be seen in Fig. 5.1. Lateral and torsional vibrations can occur inside the free span length, and these can appear to be superimposed.

The natural frequency and the deflections are influenced by the following factors:

- pre-stressing force, P_e,
- free span length, L_e,
- the asymmetric heating of the band,
- external loads.

The vibrations are always created by the effect of excitation. Excitation can originate from the cutting forces, the imperfections of the tools (uneven tooth setting or side clearance angles), the eccentricity of the band wheels and the welding of the band. It is important to understand the exciting sources because decreasing them reduces both the vibration and deflections.

The pre-stressing force is generated with different equipment on the band saws. An old, classical solution was to use tension with a dead-load, which was theoretically perfect, but its serious drawback was that it formed its own vibration system and the pre-stressing changed periodically according to the vibration. A simple solution is an external pre-stressing screw which can set a certain pre-stress, but the system does not react to changes in tension.

Further pre-stressing solutions are the spring, the hydraulic and the pneumatic pre-stressing equipment. These respond to the changes, but they do not provide a constant pre-stressing.

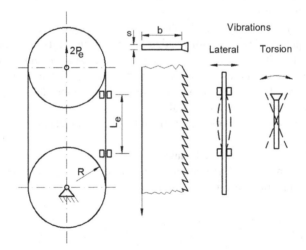

Fig. 5.1   The vibration forms of a band saw (Ulsoy and Mote 1980)

The most modern pre-stressing method is the use of a servo-motor pre-stressing equipment, which keeps the pre-stressing force at a constant value with the help of a force sensor.

The pre-set tensioning force is changed during the work by the effect of the thermal strain on the band, the unloading effect of centrifugal forces and the external forces. Therefore the effective force acting on the band can be written as follows (neglecting the external forces) (Ulsoy and Mote 1980):

$$N_x = \frac{P_e}{b} - \kappa \left(\rho \cdot s \cdot v^2 + \alpha \cdot s \cdot E \cdot \Delta \vartheta \right) \tag{5.1}$$

where:

P_e the pre-set tensioning force,
s the blade thickness,
v the blade velocity,
ϑ the temperature rise of the band,
α thermal expansion coefficient.

The κ strain system parameter characterises the wheel support method and the perfection of pre-stressing. For pre-stressing with dead-load is $\kappa = 0$, while $\kappa = 1$ for pre-stressing with screws. For spring, hydraulic and pneumatic pre-stressing methods the κ factor takes on intermediate values. The nearer κ is to zero, the more uniform the pre-stressing is.

The saw blade will run exactly in a plane given by the band wheels only if the bending stiffness of the blade is negligible. Otherwise the band deviates from the theoretical plane having a bending off-set (Wu and Mote 1984). The moment needed for the bending with radius R (Fig. 5.2) is:

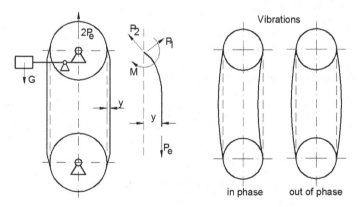

Fig. 5.2 Bending offset of a blade and vibration modes of the cutting and non-cutting spans (Wu and Mote 1984)

$$M = E \cdot I \cdot \frac{1}{R}$$

which is in equilibrium with the moment $M = P_e y$.

From the moment equilibrium, the band off-set is given by:

$$y = \frac{E \cdot I}{P_e \cdot R} \tag{5.2}$$

where $E \cdot I$ is the bending stiffness of the blade.

Depending on the wheel sizes and the pre-stressing, the value of the band off-set is 0.2–0.5 mm. The band off-set plays an important role in the vibration mechanism of the cutting and non-cutting spans. The greater the band off-set, the more it causes oscillating motion of the band wheels and the transfer of vibrations to the non-cutting span. This kind of vibration is referred to as coupling vibration (Wu and Mote 1984). The shape of vibrations may be in-phase or out phase, in some cases the superposition of the two modes occurs (Fig. 5.2). The in-phase blade vibration is always associated with band wheel oscillation.

This was observed experimentally. When the cutting span was disturbed, vibration displacements were also recorded in the non-cutting span.

There is a phase correlation of 90° between the vibrations in the two spans. Furthermore, the vibration frequency of both spans is identical and we see a high frequency vibration modulated by a low frequency oscillation. The modulation frequency depends on the tensioning (stretching) of the saw blade: increasing tension decreases the modulation frequency. This modulation phenomenon means a propagation of vibration energy between the two spans.

The stretching of the cutting and non-cutting spans is not exactly the same. There is a difference between the high frequencies in the cutting (ω_1) and non-cutting spans (ω_2). The average frequency is

$$\omega_{av} = \frac{(\omega_1 + \omega_2)}{2}$$

and the modulation frequency has a value of

$$\omega_{\text{mod}} = \frac{(\omega_1 - \omega_2)}{2}$$

and further

$$\omega_1 = \omega_{av} + \omega_{\text{mod}} \text{ and } \omega_2 = \omega_{av} - \omega_{\text{mod}}$$

The high frequency oscillations on the two spans have a 90° phase shift, i.e. one vibrates with $\cos(\omega_{av} \cdot t)$, while the other with $\sin(\omega_{av} \cdot t)$. The in-phase vibration of the blade is associated with wheel oscillations, while the out-of-phase vibration is not related to wheel oscillation.

The natural frequency of the band saw blade depends on its stretching and on stresses induced in the blade by different roll tensioning methods. Tilting of the upper wheel causes a plane bending moment in the blade which adds additional tension stresses in the region of cutting edge. Wheel crown and blade overhang on the wheel can cause similar bending moments. These axial stress components are demonstrated in Fig. 6.14 (see later).

The stress due to roll tensioning generally has values of 5000–7000 N/cm² (in Sect. 6.3) The additional stress due to tilting of the upper wheel can be approximately calculated in the following way. As a consequence of tilting, the upper edge of the wheel shifts upwards with a value ΔR and its relative value related to the wheel radius R can be expressed as

$$\frac{R + \Delta R}{R} = \cos \beta + \frac{\left(e + \frac{b}{2}\right)}{R} \cdot \tan \beta$$

where β is the tilting angle, b is the wheel width and e is the tilting eccentricity related to the centre line of the wheel; (the value of e may be zero).

Typical tilting stresses are given in Fig. 5.3 as a function of the tilting angle. The stresses increase almost linearly with the tilting angle. The combined stresses at the gullet bottom line are around 15,000 N/cm².

The natural frequency of a saw blade will be influenced by the following parameters:

The Speed of the Band It only slightly influences the natural frequency. With the increase of speed, the natural frequency decreases 5–7% in the 0–50 m/s speed range.

The Pre-stressing of the Band Pre-stressing has a determinant effect both on the transverse, and torsional vibration frequency. The transverse vibration frequency of a log rip band saw ($b = 27.5$ cm, $s = 1.5$ mm, $L_e = 60$ cm) for different pre-stressing

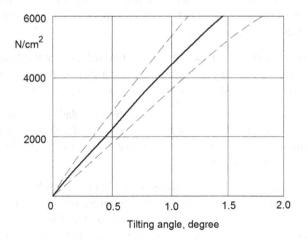

Fig. 5.3 Tilting stresses in a band saw

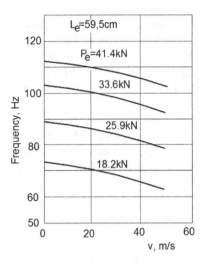

Fig. 5.4 The effect of pre-stressing of the band on its natural frequency

is shown in Fig. 5.4 (Ulsoy and Mote 1980). The Figure also shows the effect of the peripheral speed.

The Method of the Pre-stressing The method of pre-stressing may compensate the unloading of centrifugal force and temperature in different ways. An insufficient compensation decreases the natural frequency, and increases the vibration amplitudes.

The effect of the κ factor on the transverse and torsional natural frequency is shown in Fig. 5.5 depending on the peripheral speed (Ulsoy and Mote 1980).

Because a constant force wheel support system (servomotor) is expensive, it is used only on big machines. The pneumatic pre-stressing is used on small, but high quality machines, for which $\kappa \cong 0.2$. A spring-supported wheel system has values for κ around 0.5.

The Free Span Length of the Band An increasing free span length always decreases the natural frequency. The change of the natural frequencies for two free span lengths depending on the pre-stressing is shown in Fig. 5.6 (Ulsoy and Mote 1980).

An increase of the thickness and width of the saw blade decreases the natural frequency. Therefore it is wise to use thin blades. In the thin blade the bending stresses are smaller ($\sigma_m = \frac{E \cdot s}{2R}$), and therefore greater pre-stressing force can be used to increase the natural frequency.

Artificial Pre-stressing It can be induced in the blade with roll tensioning so that tensile stress will remain on the edge part and compressive stress on the midsection. Straining the band causes parabolic stress-distribution so that we get the greatest resultant tensile stress at the edges. This combined straining hardly influences the transverse vibrations, but it increases the torsional natural frequency by 10–20%.

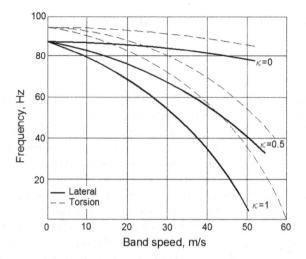

Fig. 5.5 The effect of pre-stressing methods on the change in natural frequencies

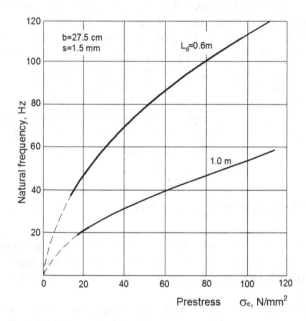

Fig. 5.6 The effect of the free span length of the band on the natural frequency

The blade heats up during the work due to friction, and a parabolic temperature distribution develops along the blade width. The exponent of the parabola changes between 2 and 3. The temperature increase causes the band to stretch in the tooth-zone, and a local decrease of the pre-stressing. This considerably lowers the torsional natural frequency.

The change in stress due to temperature variations is given by the following equation:

$$\Delta\sigma = -\alpha \cdot E \Delta\vartheta$$

where α is linear expansion coefficient (for steel $\alpha = 12 \times 10^{-6}$) and $\Delta\vartheta$ is the temperature increment. This thermal stress always decreases the local tension stress induced in the saw blade. For example, if $\Delta\vartheta = 60$ °C, then the thermal stress is around 15,000 N/cm^2, the same magnitude as the induced total tension stress in the gullet bottom line! Therefore, the thermal stresses considerably influence the natural frequency and especially the stability of saw blades.

The unloading of the edges of the blade primarily causes torsional deflections, decreasing the natural torsion frequency. The decrease depends on the temperature difference between the two edges of the blade. At 100 °C difference the torsional natural frequency decreases by 12–15%, while the transverse natural frequency hardly changes, provided that the average stress of the blade remains constant.

Furthermore, it is observed that the spectra of the blade deflection show a $1/f$ behaviour, where f is the frequency. That means that the deflection amplitudes are inversely proportional to the frequency of their occurrence.

With the tilting of the upper leading wheel, asymmetric stress distribution can be generated in the blade. The direction of the tilting causes greater than the average stresses in the tooth-zone and smaller stresses on the back edge.

The extent of the tilting can be characterized with the $(\sigma_{max}/\sigma_{min} - 1)$ value and the change of the natural frequency also depends on this. The tilting slightly increases the natural torsional frequency and decreases the natural transverse frequency in a similar extent. Therefore, the tilting of the leading wheel does not significantly reduce vibration, but improves the stability of the blade on the wheel.

5.3   The Vibrations of Circular Saws

The circular saw is one of the most frequently used woodworking machines. Owing to its simple and cheap construction and relatively great sawing capacity, its use is spreading again in log sawing (Overseas in the nineteenth century, logs were cut with circular saws).

An objection against circular saws was their high width of cut, causing higher kerf loss than other saws. This objection is no longer valid today. Reducing the amplitudes of vibrations, and improving the blade stability, thin circular saw blades can be produced now, (e.g. 0.8 mm blade thickness and 1.3 mm kerf width up to 400 mm diameter) (Berolzheimer and Best 1959).

Thin circular saw blades require the avoidance of the critical speed, the proper clamping or guiding of the blade, running the blade in its exact plane, the accurate grinding of the teeth and roll tensioning of the blade (Stakhiev 1977, 1989, 1998).

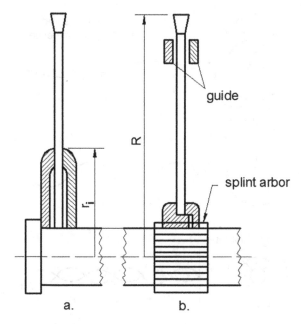

Fig. 5.7 The clamping (**a**) and guiding (**b**) of circular saws

Saw blade vibrations increase the actual kerf loss, reduce product accuracy and diminish surface quality. Therefore, efforts should be made to reduce saw blade vibrations to an appropriate level.

Today thin circular saw blades are generally held by a clamping plate. The larger the diameter of the clamping, the more stable the circular saw is, but this prevents cutting thicker logs. In the last decades different blade guiding systems have been successfully used, where two or more pairs of frictional pads guide the blade. The blade can move on its axis (Fig. 5.7), so it is self-adjusting.

The forms of vibration which develop in the rotating disk can be seen in Fig. 5.8 (Mote and Szymani 1977). The different vibration forms are characterized by the number of nodal circles and nodal diameters. With a zero nodal circle and zero nodal diameter, the total periphery of the disk bends to the right or left. With a nodal diameter of one, half of the periphery bends, and with a nodal diameter of two, a quarter of the periphery is deflected to the right or left. In larger diameter blades, a nodal circle can develop, when another wave emerges in the inner part in the opposite phase.

The vibration waves propagate along the periphery of the disk to the right and left. In a rotating disk, the travelling speed of the waves in the direction of rotation increases with the rotation speed (forward travelling wave), while the backward travelling wave speed decreases. If the frequency of the rotation parallels the frequency of the vibration, then a standing wave occurs on the disk. Here the rotation speed is called the *critical speed* (Lapin 1959).

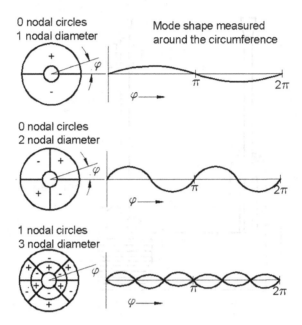

Fig. 5.8 Schematic of vibration modes of a rotating circular saw

When the operating speed approaches the critical speed, the blade becomes unstable. The large blade deflections make it impossible to produce quality sawing. The recommended maximum operating speed is at least 15% below the critical speed (Stakhiev 1977).

The frequency of the basic vibration mode can be calculated from the following expression:

$$f = \frac{\omega_0}{2\pi} = \frac{K^2 \cdot s}{4R^2 \cdot \pi} \sqrt{\frac{E}{3\rho(1 - \upsilon^2)}} \tag{5.3}$$

or using the characteristic values of steel for E, ρ and υ, we obtain

$$f = 253 \cdot \frac{K^2 s}{R^2} \tag{5.3a}$$

where the blade thickness s and the radius R must be substituted in Meters.

The constant K is a function of clamping ratio r_i/R in the following form

$$K = 1 + 5 \cdot \frac{r_i}{R}$$

For nodal diameters higher than 1, a similar equation can be used, but the constant K in Eq. (5.3a) will be changed. For clamping ratios between 0.25 and 0.3 the following constant can be used (Gogu 1997).

Nodal diameter	1	2	3	4
Constant K	253	422	717	1223

Centrifugal forces occur in a rotating disk and these forces generate tensile stresses in both radial and tangential directions. Therefore, the natural frequency of a rotating disk slightly increases as a function of rotation speed (shown in Fig. 5.9) and it can be calculated in the following manner:

$$f_\omega = \sqrt{f_n + \lambda_n \omega^2}$$

where f_n is the natural frequency at rest for a given nodal diameter n and λ_n is a constant for the nodal diameter in question. Values for λ_n are given below for clamping ratios of 0.25–0.3 (Gogu 1997).

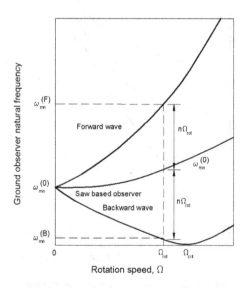

Fig. 5.9 Frequency-rotational speed diagram showing forward and backward travelling waves and the condition of standing wave resonance or critical speed

Nodal diameter	1	2	3	4
Constant, λ_n	1.25	2.3	3.75	5.5

A more general approximation is possible by using calculated nondimensional frequencies as a function of clamping ratio in the form (Mote 1965):

$$k_n = \omega_n \sqrt{\frac{4 \cdot \rho \cdot R_o^4}{E \cdot s^2}}$$

which is represented in Fig. 5.10.

From the above equation, using the appropriate material constants, we get

$$\omega_n = 2590 \cdot k_n \frac{s}{R_o^2} \tag{5.4}$$

where the blade thickness s and the blade radius R_o must be substituted in Meter. For example: with $R_o = 250$ mm, $s = 2$ mm and $R_c/R_o = 0.4$, the value of k_n from the Figure

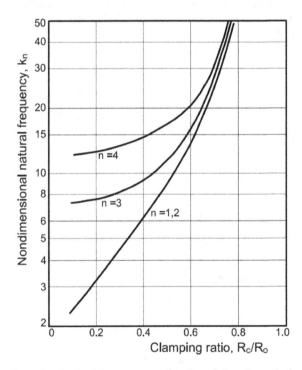

Fig. 5.10 Non dimensional natural frequency as a function of clamping ratio for different nodal diameters (replotted from Mote 1965)

$$n = 2 \; k_n = 6.4 \; \omega_n = 530 \text{ rad/s} \quad f_n = 84 \text{ Hz}$$
$$\text{for} \quad n = 3 \; k_n = 9.4 \; \omega_n = 779 \text{ rad/s} \quad f_n = 124 \text{ Hz}$$
$$n = 4 \; k_n = 15 \; \omega_n = 1243 \text{ rad/s} \; f_n = 198 \text{ Hz}$$

Increasing the clamping ratio shifts all natural frequencies upward as seen in Fig. 5.10. From the standpoint of stability it is essential that the free radius $(R_o - R_c)$ does not significantly exceed the cutting height. Working with thin saw blades, it is important that the clamping ratio and blade radius would be adjusted to maximize the natural frequency corresponding to critical speed for the desired cutting height.

The above frequencies are the "true natural frequencies," which could be seen by an observer at the saw.

The vibration waves propagate in opposite angular directions around the saw blade. Therefore, a ground based observer (or at the position of the workpiece) sees forward and backward travelling waves corresponding to a given mode n:

$$\omega_{n1} = \omega_n + n \cdot \Omega \quad \text{forward wave}$$
$$\omega_{n2} = \omega_n - n \cdot \Omega \quad \text{backward wave}$$

where Ω means the angular velocity of the saw blade, ω_n is the angular speed corresponding to the natural frequency.

The blade rotation decreases the angular speed of the backward wave and, therefore, ω_{n2} can approach the value of zero. This is the critical speed of the blade and the ground based observer sees a standing wave (Lapin 1959). If a standing wave occurs, the external forces always excite the blade edge at the same point which leads to an accumulation of energy causing instability (large deflections) of the blade. Therefore, the operating speed of the saw must be reduced at least 15% below the critical speed (Stakhiev 1977).

If the blade has edge slots a standing wave cannot be established at the rim of a blade and this reduces the amplitude of the unstable motion. This means, resonance and critical speed can still occur, but the slots inhibit the formation of standing waves (Mote and Szymani 1977).

Stresses induced in a disc originate from centrifugal forces, uneven temperature distribution and from introduction of tensioning stresses (see Sect. 6.3).

The common distribution of these membrane stresses in radial and tangential directions is shown in Fig. 5.11 (Mote and Szymani 1977).

The rotational forces depend on the radius, density of the material and on the squared angular speed. The radial and tangential stresses induced in the disc also depend on the clamping ratio. The typical stress distribution can be seen in Fig. 5.12. The radial stress increases from zero on the outer edge towards inside depending on the clamping ratio.

The tangential stress at the rim is not zero and its value increases as the clamping ratio decreases. It is very important that the rotational stresses are always tensile and contribute to the stability of the blade.

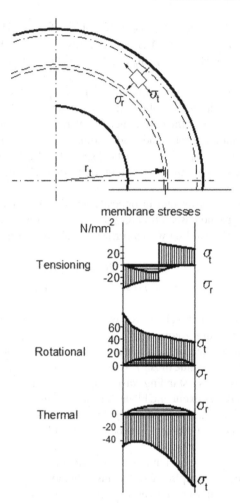

Fig. 5.11 Membrane stresses in a circular saw

The thermal stresses depend on the coefficient of thermal expansion and the temperature distribution induced by sawing. The heat generated by friction occurs on the cutting edge and on both side of the gullet face so the disc temperature generally increases towards the outer edge in the following general form:

$$\vartheta(r) = \vartheta_0 + \Delta\vartheta \left(\frac{r}{R}\right)^n$$

where r/R means the relative radius of the sawblade, ϑ_0 is the ambient temperature and the exponent n varies between 2 and 3.

The radial thermal stresses increase almost linearly towards the centre of rotation, while the tangential stresses have their maximum value along the rim of the disc

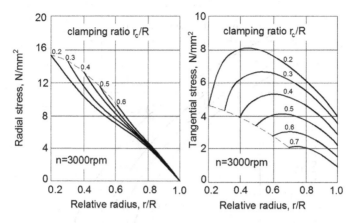

Fig. 5.12 Rotational stresses in a circular saw (Pahlitsch und Friebe 1973)

(Fig. 5.13). The radial stresses are again tensile, but the large tangential stresses are compressional which may cause a corrugated edge and the disc loses its stability. Therefore, excessive thermal gradients in the disc must be avoided.

Tensioning stresses, generally induced by rolling (see Sect. 6.3), should decrease or prevent excessive tangential stresses which would cause a lower natural frequency and make the disc unstable.

The combined effect of thermal stresses and rotational stresses is demonstrated in Fig. 5.14 (Gogu 1997). Temperature differences strongly lower the natural frequency for $n \geq 2$ nodal diameters. At the same time, an increasing rotational speed increases the allowable temperature difference.

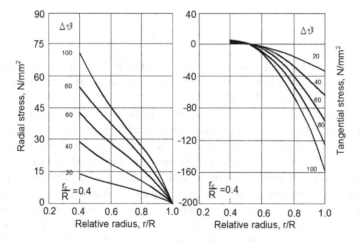

Fig. 5.13 Thermal stresses in a circular saw (Pahlitsch und Friebe 1973)

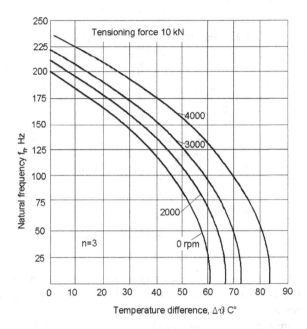

Fig. 5.14 Effect of temperature differences and rotational speed on the third natural frequency for a circular saw having $D = 400$ mm and clamping ratio of 0.3 (Gogu 1997)

Experimental measurements on a circular saw 500 mm in diameter and 2.5 mm thick are shown in Fig. 5.15 (Stakhiev 1977). The critical temperature difference, at which the natural frequency tends toward zero, is around 50 °C. The critical temperature difference can be estimated by the following equation (Gogu 1997):

$$\Delta \vartheta_{cr} = A \frac{1}{\alpha} \left(\frac{s}{R} \right)^2$$

where α is the thermal expansion coefficient, s is the thickness and R is the radius of the saw. The constant A is a function of clamping ratio and may be substituted in the range of 0.25–0.3, as $A = 5.0$–5.5.

Theoretical and experimental investigations in the 1950s have clearly shown that thermal stresses in circular saws increase the natural frequency for the modes n = 0 and n = 1 but decrease for the modes n ≥ 2. The effect of roll tensioning is opposite to that of thermal stresses. Therefore, the unfavourable action of thermal stresses can fully be compensated by the use of artificial prestressing (Zhodzisskiy 1958). The effect of roll tensioning on the natural frequencies is shown in Fig. 5.16 (Gogu 1997). Generally the influence of tensioning is higher for small clamping ratios. For a given value of the rolling force, for the modes $n = 0$ and $n = 1$, the saw loses its elastic stability and buckling (dishing) occurs. That is the *critical rolling force*, at which the disc is no longer flat but forms a dish. The optimum rolling radius, for the highest possible natural frequency, is generally around 0.7R.

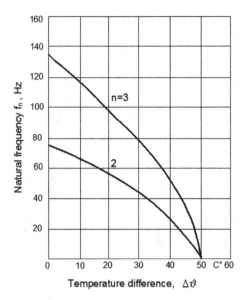

Fig. 5.15 Effect of temperature differences on the second and third natural frequencies of a circular saw with $D = 500$ mm and s $= 2.5$ mm (Stakhiev 1977)

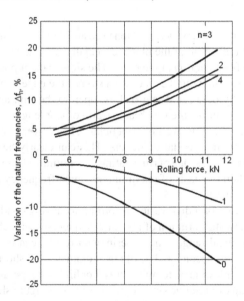

Fig. 5.16 Variation of natural frequencies as a function of tensioning force (Gogu 1997)

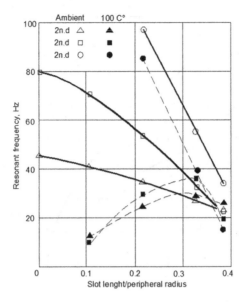

Fig. 5.17 The effects of slot length on the resonant frequencies with rim temperatures at ambient and 100 °C above ambient. Disk 460 mm in diameter, 2.4 mm thick at 2100 rpm

Slots at the edge of circular saws are often used and they have a significant effect to decrease the lateral deflection of the blade, especially when the saw blade subjected to excess heat load from friction at the edge. The slots make it possible to eliminate the tangential stresses allowing the free expansion of the edge zone.

There are generally between 4 and 6 slots, open to the rim, and the length of a slot is related to the radius of the saw blade, (ratio of slot length).

Without rim heating, slots are not beneficial. They lower resonant frequencies in modes with two or more nodal diameters. This can clearly be seen in Fig. 5.17 (Mckenzie 1973). In the presence of rim heating, an optimum slot length ratio at about 0.3 was determined on the basis of maximum resonant frequency in the second and third nodal diameter mode.

The rim temperature of a disk varies according to sawing conditions. Therefore, it is important to know the variation of resonant frequencies with changes of temperature within the practical range. The experimental results can be seen in Fig. 5.18 (Mckenzie 1973) for the second and third nodal diameters. A short slot gives higher resonant frequency for lower rim temperatures. For higher temperatures, a longer slot gave better results up to a relative slot length of 0.33. The Figure shows the vibration frequency equal to the rotation speed (35 Hz = 2100 rpm) by a horizontal dotted line with intersections giving resonant vibrations. These experimental results show, that the low resonant frequencies do not allow higher rotation speeds, especially with longer slots.

The most hazardous situation to be avoided occurs when the rim temperature reaches a critical value, at which the resonant frequency falls near zero. The disk

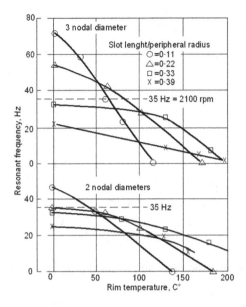

Fig. 5.18 The effect of slot length and rim temperature on resonant frequencies for a disk rotating at 2100 rpm, five slots. Disk 460 mm in diameter, 2.4 mm thick

lacks any stiffness against lateral forces. In practice, the slotting of circular saws has been mainly used on carbide-tipped saws operating at lower temperatures. It may be better to make the slots shorter to allow room inside them for tensioning.

It was generally believed that closed slots in a disk have little effect on the operating conditions. Recently, a special closed slotting was published (Satoru 2005) allowing a thinner blade and a smaller kerf. The side view of the saw blade with special S-shaped slots is shown in Fig. 5.19. Measurements were carried out with this blade and the critical rotation speed and the critical rim temperature increased up to 60% compared to a conventional blade.

Another interesting approach is the use of guided splined arbor circular saws (Fig. 5.20) operating at rotational speeds above the lowest critical speed (Lister et al. 1997). This supercritical speed sawing has the potential to reduce saw kerf thickness while maintaining good sawing accuracy and feed speed. Saws with 1.5 mm kerf were used to cut 100 mm lumber at 40 m/min feed speed with sawing deviation of less than 0.25 mm.

Figure 5.21 shows the critical speed variation of a guided saw using different blade thicknesses. In practice, the saws are run between the second and third critical speeds. The speeds should be selected where the idling saws indicate the best stability.

The sawing performance is important in practice because it illustrates the relationships among feed speed, saw blade thickness and sawing accuracy. This relationship is given in Fig. 5.22 (Lister et al. 1997). The results in the Figure clearly show the effect of both feed speed and saw plate thickness on the sawing accuracy. As could be expected, the stiffest 1.5 mm thick saw has the best sawing accuracy and can be

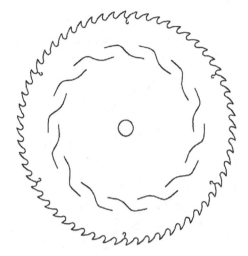

Fig. 5.19 Side view of stable saw blade with *S*-shaped slots (Satoru 2005)

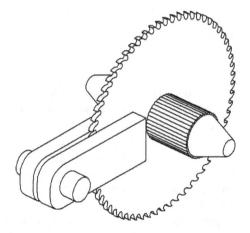

Fig. 5.20 Saw using splined arbor with guides (Lister et al. 1997)

used with a high feed speed. The 1.0 mm thin blade has higher sawing deviations and requires a lower feed speed. The maximum feed per tooth for the blade thicknesses shown in Fig. 5.22 are 0.65 mm, 0.85 mm and 1.0 mm respectively.

Another interesting development was the "hyper critical saw blade" with centre slots to improve the stability of a thin kerf saw blade and its resistance to thermal buckling. The idea of this slot spacing was to allow the free radial expansion of the blade from centrifugal forces of rotation (Fig. 5.23).

This saw blade is also fixed by flanges from both sides. However, tightening the flange will limit the free expansion of the blade and above a certain torque, the saw blade would work as a conventional saw blade. Therefore, the flange is tightened

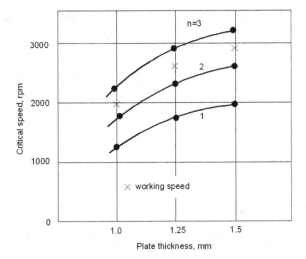

Fig. 5.21 Critical speed variation as a function of blade thickness for a guided saw 500 mm in diameter (Lister et al. 1997)

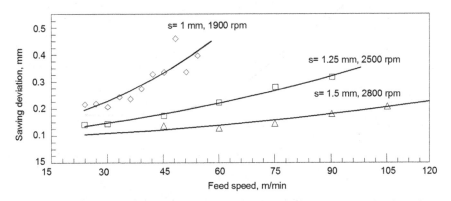

Fig. 5.22 Sawing deviation versus feed speed for a guided saw using different blade thicknesses. Cutting height is 100 mm, with 42 teeth

with only enough torque to keep the saw blade stiff. The effect of free expansion on the critical rotational speed can be seen in Fig. 5.24 for a saw blade 305 mm in diameter, a 1.5 mm kerf and 0.9 mm thick blade. The Figure shows that with free expansion, the critical rotational speed is at least 50% higher compared to a rigidly fixed flange. This saw blade was used to cut wood 60 mm thick at a rotation speed of 4000 rpm, a feed speed of 80 m/min and a 0.5 mm feed per tooth.

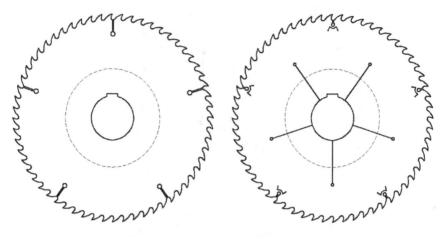

Fig. 5.23 Conventional (left) and hyper critical saw blade (right) (Satoru 2003)

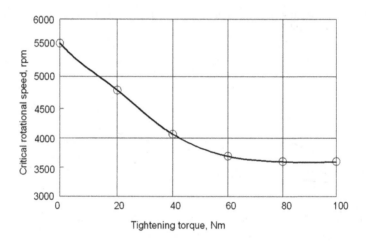

Fig. 5.24 Relationship between tightening torque and critical speed (Satoru 2003)

In the development of circular saws, all natural frequencies are calculated and experimentally confirmed. These results give the range of critical rotational speeds which must be avoided on both sides, at least, to a distance of 15% from the critical values. The calculations and experimental results are shown in Fig. 5.25 (Mote and Szymani 1977).

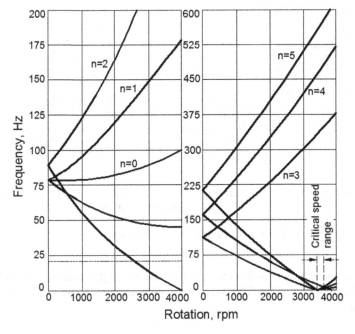

Fig. 5.25 Frequency-rotational speed diagrams showing the critical speed range

5.4 Washboarding

Band saws and circular saws often cause a wave like surface pattern on surface of sawn lumber which is widely known as washboarding. This phenomenon occurs when using higher blade speeds, faster feed speeds and thinner band and circular saws. A washboarding profile can be up to 0.6–0.8 mm deep on the face of the lumber. Therefore, the rough size must be large enough to remove sawing deviations and the depth of washboard. This considerably decreases lumber recovery.

Washboarding is a dynamic phenomenon occurring at high frequencies. It was observed that the main cause of washboarding is the self-excited vibration during sawing (Okay et al. 1995; Lehmann and Hutton 1997; Orlowski and Wasielewski 2001). Furthermore, an important condition causing washboarding is when the frequency of the tooth passage is slightly higher than its natural frequency. It is also concluded that all the washboards for band saws were produced by torsional vibration.

The typical wave like pattern of a washboard is shown in Fig. 5.26 (Okay et al. 1995).

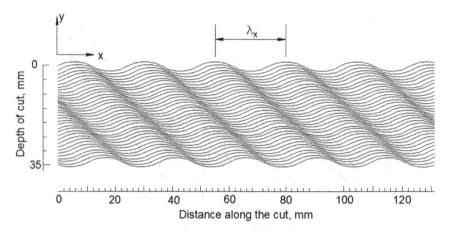

Fig. 5.26 Typical wave-like pattern of a washboarding band saw

The wave-like pattern is characterized by the wave length in feed direction (λ_x) and in the direction of cut (λ_y). The lateral tooth displacement is given by the harmonic equation

$$y = y_0 \sin(2\pi \cdot f_n t)$$

where y_0 is the maximum lateral displacement of a tooth due to self-excited torsional vibration with natural frequency f_n.

The wave length of this vibration is

$$\lambda = \lambda_y = \frac{v}{f_n}$$

where v is the band speed. The tooth frequency f_t for a given point is calculated as

$$f_t = \frac{v}{t}$$

where t means the tooth pitch. If $f_t = f_n$, then each tooth approaches a given point in the same phase. Therefore, peaks and valleys run parallel to the feed direction.

Experiments have shown that a fully developed washboarding always occurs at tooth passing frequencies that are slightly higher than the natural frequency of the blade (Fig. 5.27) (Okay et al. 1995).

The band natural frequency was 196 Hz and washboarding occurred between tooth frequencies of 196 and 204 Hz. This is a resonance phenomenon and energy will be accumulated in the subsequent cycles, which increases the lateral displacement. It may be assumed that the slight difference in the frequencies will increase the excitation forces. When there is a slight difference between the two frequencies, a

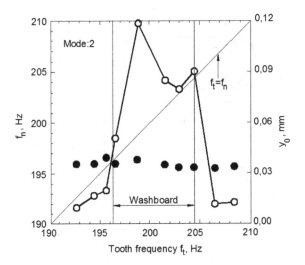

Fig. 5.27 Washboard range just after a resonance (black points show resonance frequency)

small ΔT time difference occurs between the period of tooth passage and the period of band saw vibration (Fig. 5.28) (Okay et al. 1995).

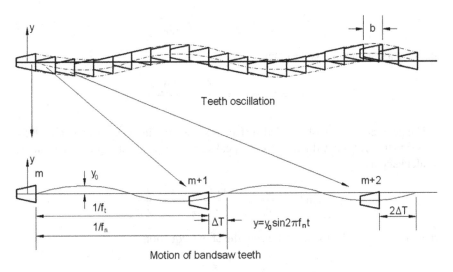

Fig. 5.28 Time difference between tooth passing frequency and blade resonant frequency

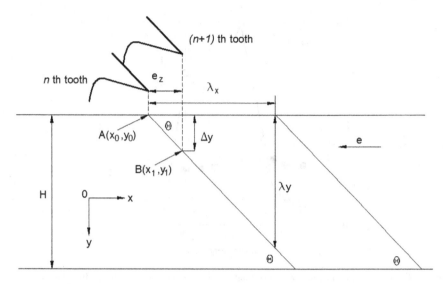

Fig. 5.29 Geometry of a washboarded surface of a workpiece

This time difference is given by

$$\Delta T = \frac{1}{f_t} - \frac{1}{f_n} = \frac{-(f_t - f_n)}{f_t \cdot f_n}$$

and the corresponding displacement

$$\Delta y = v \cdot \Delta T = \frac{-(f_t - f_n) \cdot t}{f_n}$$

The geometry of a washboarded surface is shown in Fig. 5.29 (Okay et al. 1995).
Due to the frequency differences, the peaks and valleys are at angle Θ. This angle is calculated as

$$\tan \Theta = \frac{\Delta y}{e_z} = \frac{-v(f_t - f_n)}{e \cdot f_n} = \frac{\lambda_y}{\lambda_x}$$

where $e_z = e/f_t$ is the tooth feed. From the above equation

$$\lambda_x = \frac{e}{f_t - f_n}$$

If $f_t = f_n$, then $\lambda_x = \infty$ and that means that the peaks and valleys on the surface run parallel to the feed direction.

To minimize the effect of washboarding the following suggestions can be done:

Because washboarding usually occurs in a specific rotation speed range, washboarding may be eliminated by a 5–10% increase or decrease of the blade speed. Somewhat similar effect can be achieved by increasing or decreasing of the band saw tension, which alters the natural frequency of the saw blade. A thinner blade has a lower natural frequency and is more likely to cause washboarding. Thus a thicker blade may reduce or eliminate washboarding.

Excitation forces depend on the tooth bite which could be reduced by reducing the tooth pitch.

5.5 The Vibrations of the Workpiece

The machined workpiece forms a vibration system and causes vibration displacements while running through the machine. The vibration displacements worsen the accuracy of the processing and the surface quality, so vibration deflections have to be kept as low as possible.

The vibrations of the workpiece depend on many factors. The most important are:

- the thickness and length of the workpiece,
- the exciting forces and the instantaneous boundary conditions in relation to the location of machining.

Boundary conditions mean the position of the pressing rollers, their pre-stressing and the spring stiffness.

These experimental results are from a multi-head moulder, but the relationships refer implicitly to thickness planer and shaper moulders and milling machines (Sitkei et al. 1990).

The principle of a multi-head moulder and its mechanical model can be seen in Fig. 5.30. The machined material is pressed to the machine table by spring loaded rollers. The vibration mass is represented by the machined material, the spring constants on the two sides of the mass differ from each other by at least two orders of magnitude. The machine table can be considered rigid in relation to the workpiece. Therefore, the spring constant on the side of the machine table is determined by the elasticity of the wood. This spring constant primarily depends on the thickness of the workpiece related to the elastic modulus (H/E). On the upper side of the wood the compressive force is determined by the pre-stressing of the rollers and their spring characteristics. The frequency of the exciting forces is usually between 400–800 Hz. They depend on the tool rotation speed and the number of the knives, the mechanical system of the vibrating mass and the change of the forces depending on their displacement are shown in Fig. 5.31. The forces on the machine table increase very sharply. The forces on the side of the rollers vary moderately. We conclude that the vibration deflections will be highly asymmetric.

The possible vibration modes of a workpiece are shown in Fig. 5.32. They can have transverse, rotating (torsional) and bending vibrations. The first two vibration forms are characteristic of shorter and thicker workpieces. Bending vibrations are

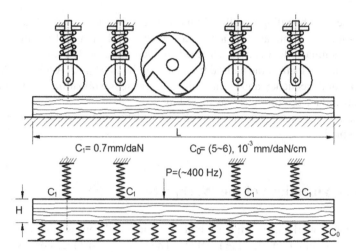

$C_1 = 0.7$ mm/daN $C_0 = (5\sim6)$, 10^{-3} mm/daN/cm

$P = (\sim400$ Hz)

Fig. 5.30 Sketch of a multi-head moulder and its mechanical model

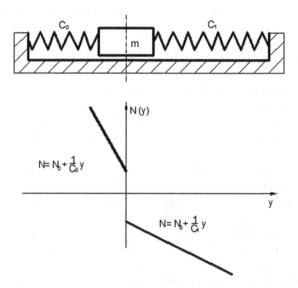

$N = N_0 + \dfrac{1}{C_0} y$

$N(y)$

$N = N_0 + \dfrac{1}{C_1} y$

y

Fig. 5.31 The forces acting on the vibrating mass depending on its displacement

typical for the longer and thinner workpieces. Rotating vibration can appear at the entry and exit of the workpiece into and out of the machine, since at this time the force pressing down the workpiece is asymmetric.

The nodes of the bending vibration waves only are under the pressing rollers, if the spring constant of the pressing rollers is big enough providing a rigid clamping. The stiffness of the pressing rollers can be characterized with the following, dimensionless number:

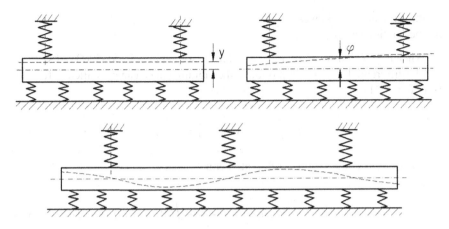

Fig. 5.32   The possible vibration modes of a workpiece

$$K = \frac{c_1 \cdot E \cdot I}{l^3}$$

where:

c_1 stiffness of the spring loaded rollers,
EI bending stiffness of the machined board,
l the distance of the pressing rollers.

For multi-head moulders, K values change between 4 and 5, and this means that the frequency of the vibration depends on the length of the workpiece, not on the distance between the pressing rollers. Pressing by the rollers does not mean rigid clamping.

The exciting force is given primarily by the radial component of the cutting force. The radial component depends on the sharpness of the knife, the chip thickness and the mechanical properties of the wood. The exciting forces per single rotation for different numbers of teeth are shown in Fig. 5.33. The curve of the exciting forces can be expressed with the Fourier series expression (or FFT analysis). If we assume that the exciting force agrees with a half period ($z = 8$–12), then we get the following infinite series:

$$P(t) = P_0 \left(\frac{1}{\pi} + \frac{1}{2} \sin \omega \cdot t - \frac{2}{3\pi} \cos 2 \cdot \omega \cdot t - \frac{2}{15\pi} \cos 4 \cdot \omega \cdot t - ... \right) \quad (5.5)$$

The angular frequency of the exciting forces:

$$\omega = \frac{z \cdot n}{9.55}$$

where:

z the number of the knives,
n the rotation speed.

When $n = 6000$ rpm and for $z = 4–8$ knives we get $\omega = 2500–5000$ rad/s.

The differential equation of motion of the vibrating mass can be written as follows (see: Fig. 5.31):

$$m\frac{d^2y}{dt^2} + \left(N_0 + \frac{y}{c}\right) = P(t) \tag{5.6}$$

The general solution of the differential equation is:

$$y = A_1 \cos \alpha \cdot t + A_2 \sin \alpha \cdot t + \frac{1}{m \cdot \alpha} \int_0^t P(t) \cdot \sin(t - \tau)d\tau \tag{5.7}$$

and $\alpha = \frac{1}{\sqrt{m \cdot c}}$.

where:

A_1, A_2 integration constants,
α the angular frequency of the free vibration.

The first two parts of Eq. (5.7) describe the free vibration of the system, while the third part gives the forced vibration. The vibration is very asymmetric because the spring constants are very different on the two sides of the mass and they are not coupled to each other. We can see the developing vibration picture in Fig. 5.34, which is strongly asymmetric, and the forced vibration is superimposed on the free vibration with a substantially larger frequency. The following relation exists between

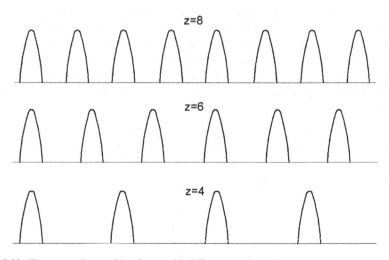

Fig. 5.33 The succeeding exciting forces with different numbers of teeth

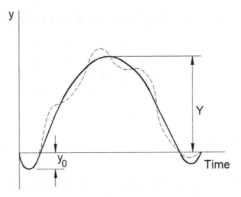

Fig. 5.34 The asymmetric vibration of the workpiece

the two amplitudes:

$$Y = \frac{\alpha_0}{\alpha_1}(y_0 - N_0 \cdot c_0)$$

where:

α_0 angular frequency of the workpiece against the machine table,
α_1 angular frequency of the workpiece against the spring loaded rollers.

It can be seen from the equation that the Y amplitude can be decreased substantially by pre-stressing the supporting rollers (Fig. 5.35).

The amplitude of the forced vibration results from the integration of the third part of Eq. (5.7) If the $P(t)$ disturbing function is given with Eq. (5.5), then the displacement given by:

$$y = \frac{P_0}{m \cdot \alpha^2}\left[\frac{1}{\pi} + \frac{1}{1 - \left(\frac{\omega}{\alpha}\right)^2} \cdot \frac{1}{2}\sin\omega \cdot t - \frac{1}{1 - \left(\frac{2\omega}{\alpha}\right)^2} \cdot \frac{2}{3}\cos 2\omega \cdot t - \ldots\right] \quad (5.8)$$

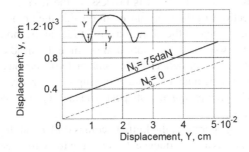

Fig. 5.35 The effect of the pre-stressing on vibration deflection (Sitkei et al. 1990)

The angular frequency of the bending vibration can be calculated from the following equation:

$$\alpha_b \cong 9\sqrt{\frac{E \cdot I}{q \cdot L^4}} \tag{5.9}$$

where:

q the mass per unit length,
L the length of the rod.

The results of these calculations are summarized in the following table, which refer to a 4 × 5 × 150 cm window frame component.

The values of the vibration parameters:

Parameters	Free vibration	Forced vibration
α_0	12,780 rad/s	
α_1	184.6 rad/s	
α_b	208.8 rad/s	
ω	–	3770 rad/s
y_0	4–6 μm	
Y_{free}	110–240 μm	
Y_{for}	–	50–70 μm

Acceleration sensors were mounted on this workpiece, and the vibration amplitudes of the workpiece were measured during the processing cycle.

The vibration amplitudes are very small toward the machine table, but their frequency (α_0) is large. On the contrary, the vibration amplitudes toward the supporting rollers are much higher. It is interesting to note that the calculated transverse vibrations (α_1) and the angular frequency of the bending vibrations (α_b) do not substantially deviate from each other.

The measured vibration amplitudes of the workpiece along its length are shown in Fig. 5.36. When the moulder-head is processing the front and the back end of workpiece, these sections of the workpiece fall outside the immediate effect of the pressing rollers. That increases the vibration amplitudes at the moulder-head. When the rollers support the workpiece in relation to the moulder-head centrally, then the deflections are minimized. The deflections depend on the direction of exciting forces, whether they act toward the machine table or the supporting rollers. In the latter case the deflections will be greater (e.g. in a shaper). The measured deflections are similar to the calculated values.

During the working operation, the knife becomes duller and the radial force will continuously increase. As a consequence, the exciting force increases and the deflections will also increase. This can be seen clearly for the shaper and thickness planer in Fig. 5.37.

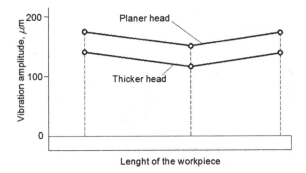

Fig. 5.36 Vibration amplitudes generated by thickness planer and shaper along the length of the workpiece (Sitkei et al. 1990)

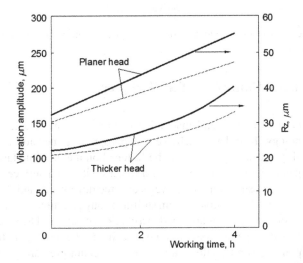

Fig. 5.37 The effect of working hours on vibration amplitudes and surface roughness (Sitkei et al. 1990)

The periodic transverse deflection of the workpiece has a substantial effect on its surface quality. We have illustrated the surface roughness together with the deflections in Fig. 5.37 (continuous lines). The surface roughness increases similar to the vibration amplitudes.

Further measurements have shown that there is a close correlation between the vibration amplitudes and the surface quality. The correlation can be seen in Fig. 5.38. A dull knife produces a slightly worse surface quality at the same amplitude, as the edge of its larger radius causes more damage to the surface. But the definitive factor is always the vibration amplitude. The surface quality is not constant along the length of longer workpieces. It is always the best in the middle parts, where the vibration deflections are the smallest.

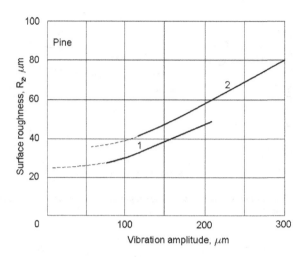

Fig. 5.38 Influence of vibration amplitude of the workpiece on the surface roughness. 1—sharp tool; 2—blunt tool

5.6   Pneumatic Clamping of a Workpiece

Computer controlled routers and machining centres represent the leading wood-working technology. In order to achieve an appropriate dimension accuracy and maximum quality, the workpiece has to be fastened on the machine table securely. The most frequent clamping method for workpieces with a flat surface is a vacuum field. The vacuum pressure generates a vertical force downward, and the horizontal friction force prevents the workpiece from shifting in any direction. The surface of the clamping table contains grooves arranged in a checkered pattern. The vacuum area of the desired size and shape may be formed by placing rubber profiles in the grooves. The vacuum pump generally allows a maximum vacuum pressure of 0.8–0.9 bar. The pressing force generated by the vacuum is

$$F_v = p_v \cdot A$$

where p_v is the vacuum pressure and A is the effective area under pressure.

A workpiece clamped by vacuum on the machine table can be shifted by applying a horizontal force. Figure 5.39 shows the force and displacement relationships for a given workpiece (Csanády and Németh 2005).

The horizontal counter force on the workpiece is a friction force and, so the friction coefficient can be calculated:

$$F_h = \mu_0 \cdot F_v \text{ or } F_h = \mu \cdot F_v$$

where μ_0 is the static friction coefficient for a resting body, and μ is the kinetic friction coefficient for a body is in motion.

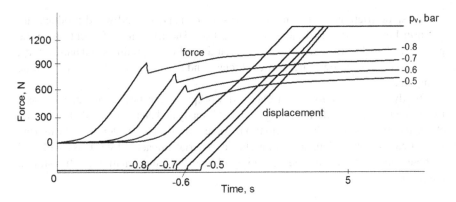

Fig. 5.39 Force requirement for shifting a workpiece on the machine table. Laminated particle board, $A = 583$ cm^2, at different vacuum values

In Fig. 5.39 we can see that the static friction coefficient is slightly higher than the kinetic friction coefficient. After a given sliding distance, the kinetic friction coefficient slightly increases and it can surpass the static value. From a practical point of view, the static friction coefficient is definitive.

To calculate the friction coefficient, the vertical pressing force must be known. Apparently this is easily calculated from an equation given above, but the selection of the effective surface area under suction may cause some troubles. Measurements have shown that the ratio of effective and actual surface area depends on the shape and size of the workpiece and also on the strength of the vacuum. Generally this ratio varies between 0.80 and 0.90 and the higher values are valid for larger workpieces and lower vacuum pressures (Csanády and Németh 2005). For MDF, however, lower values between 0.7 and 0.8 are obtained (Fig. 5.40).

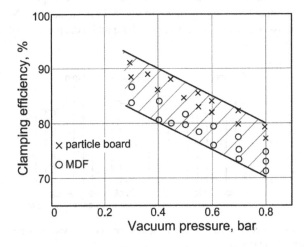

Fig. 5.40 Relationship between vacuum pressure and clamping efficiency

The static friction coefficients for various wood species and wood products are given in Fig. 5.41 (Csanády and Németh 2005). The friction coefficient along the grain is slightly less than across the grain. Laminated wood products and hard woods, such as oak, have a lower friction coefficient and are more sensitive to inaccurate clamping.

The dynamic behaviour of a clamped workpiece can be examined by impact forces made by a pendulum loading device (Csanády and Németh 2005). The impact characteristics such as the deformation, force, speed and deceleration during the impact can also be calculated as an elastic material impacted by a mass with and without sliding of the workpiece. Taking a linear force deformation relationship in the form

$$F = B \cdot x$$

where B means the stiffness of the material, then the maximum deformation is given by the energy equation (see Fig. 5.42):

$$M \cdot g \cdot h = \int_0^x F \cdot dx \text{ and } x_{max} = \sqrt{\frac{2 \cdot M \cdot g \cdot h}{B}} \qquad (5.10)$$

$$F_{max} = B \cdot x_{max} = \sqrt{2B \cdot M \cdot g \cdot h}$$

The above equations are valid if the counter-force is greater than the maximum impact force. If not the impacted body will slide to a given distance and the surplus energy of the mass will be consumed by friction. It is important to note that a considerable part of the impact energy is converted into deformation and acceleration of the body. Theoretically, a pure elastic deformation does not consume energy and the impacting mass would be returned to its initial position. The displacement of an oak specimen during impact loading is shown in Fig. 5.43 (Csanády and Németh 2005). The impacting mass was $10 \times 10 \times 10$ cm, 7.8 kg and a drop height of 7.5 cm. The

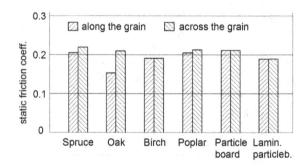

Fig. 5.41 Static friction coefficients along and across the grain for various wood materials in a vacuum range from 0.7 to 0.8 bar

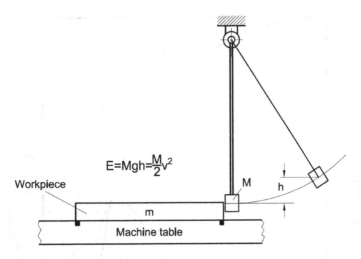

Fig. 5.42 Pendulum loading device (Csanády and Németh 2005)

impact energy is 5.74 N m and the impact speed is 1.21 m/s. The friction force of the specimen was appr. 1500 N. Friction energy of 2.83 N m was consumed over a sliding distance of 1.83 mm and this amount is only 49% of the total impact energy.

The energy of impact is consumed by the friction force at sliding and by the acceleration of workpiece. The energy balance equation yields

$$\frac{M}{2}v_0^2 = Mgh = \mu \cdot N \cdot y_0 + F_a \cdot \frac{y}{2} \text{ with } v_0 = \sqrt{2gh}$$

where

μN is the friction force,
y is the displacement during impact and acceleration,
y_0 is the sliding distance of the workpiece.

The inertia force F_a is approximately calculated as

$$F_a = m \cdot a = m\frac{2 \cdot y}{t_y^2} = m\frac{x \cdot 2 \cdot y_0}{t_y^2} \text{ with } y = x \cdot y_0$$

Using the above approximation, the energy consumed by the acceleration is calculated as

$$F_a \cdot \frac{y}{2} = m\frac{x^2 \cdot y_0^2}{t_y^2}$$

where t_y means the duration of acceleration (see in Fig. 5.43).

Keeping in mind the above relationships, the energy equation has the form:

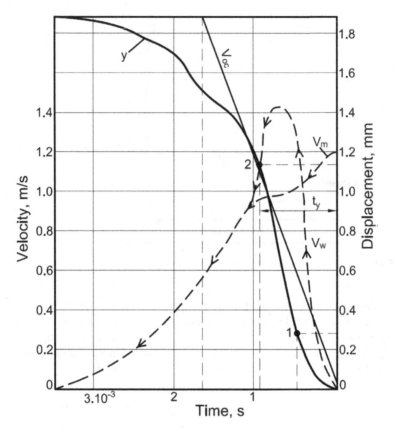

Fig. 5.43 Unsteady motion of workpiece after impact load,$h = 7.5$ cm, $m = 1.9$ kg, vacuum is—0.8 bar, $\mu.N = 1500$ N

$$\frac{m \cdot x^2}{t_y^2} \cdot y_0^2 + \mu \cdot N \cdot y_0 - Mgh = 0 \qquad (5.11)$$

Figure 5.43 clearly shows the effect of inertia forces after the approaching of the pendulum mass. The velocity of workpiece gradually increases and to the end of true acceleration by the mass M (point 1) the velocity of workpiece more or less equals the theoretical common velocity. Thereafter the stored elastic energy further accelerates the workpiece and, the end of acceleration (point 2), the velocity of workpiece nearly corresponds to the impact velocity v_0. In the following, the workpiece decelerates due to friction forces.

The stored elastic deformation, during its rebound, covers some portion the energy of acceleration and therefore, the first term in Eq. (5.11) is actually smaller. The energy balance equation has the following form:

$$\psi \frac{m \cdot x^2}{t_y^2} y_o^2 + \mu \cdot N \cdot y_o - M \cdot g \cdot h = 0 \text{ and } \Psi < 1 \qquad (5.11a)$$

which can be solved for the displacement y_0.

In the experiment shown in Fig. 5.43 the following values can be read: $x = 0.6$, $t_y = 0.8 \times 10^{-3}$ s, and using these value with $\psi = 0.7$, the deformation is $y_0 = 1.96$ mm, $y = x.y_0 = 1.15$ mm with good agreement with the measured values. The value of x is, however, not known in advance. Measurements results gave values for x between 0.5 and 0.6 and, therefore, a similar guess value gives acceptable results. The Figure shows the velocity of impact mass and workpiece during the impact and common sliding.

It is interesting to note that the average accelerations calculated to the points 1 and 2 are nearly the same (3500 m/s^2). Therefore, the use of point 2 to calculate the inertia force F_a causes no considerable error. At the same time, the location of point 2 is more clearly indicated.

The specimen on the machine table is also supported by the elastic rubber which exerts an additional resistance only after a given deformation. Therefore, the displacement curve shows a wavy character.

Softwood species, and particle board, suffer more deformation and absorb more energy. This means less surplus energy and less displacement, which was verified experimentally (Fig. 5.44).

If the impact force is less than the friction force then no sliding occurs. Bigger impact forces due to machining may, however, cause the vibration of the workpice.

The pneumatic clamping of a workpiece cannot be regarded as a rigid clamping method and the workpiece is often inclined to vibrate. Especially small workpieces

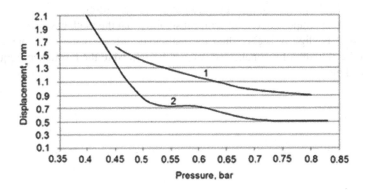

Fig. 5.44 Displacement of oak (1) and chipboard (2) specimens as a function of clamping pressure. Impact mass 7.8 kg, drop height 7.5 cm, clamping vacuum 0.6 bar, $A = 583$ cm^2 (Csanády and Németh 2005)

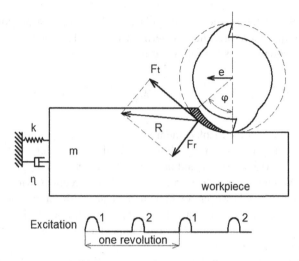

Fig. 5.45 Excitation of a workpiece by periodic cutting forces

tend to have forced vibrations due to the excitation of the cutting force (Fig. 5.45). The governing equation of motion is similar to Eq. (5.6) with the difference that here also a damping term ($\eta \cdot \dot{y}$) should be included. Resonance occurs if the natural frequency of the workpiece equals the force excitation frequency or its multiple. A typical feature of this phenomenon is the unequal tooth feed and cutting force on the individual cutting edges, as shown in Fig. 5.46.

In the first case the feed per revolution is large enough ($e_n = 1.67$ mm) and the displacement of the workpiece in the same direction to the approaching cutting edge can only halve the tooth feed. In the second case the feed per revolution is 0.83 mm and the second cutting edge has practically no tooth feed. The edge is in contact with the workpiece which is indicated by the radial force. To avoid resonance, the frequency of rotation should be changed. If one cutting edge fails to cut, the surface will generally show a wavy character similar to washboarding.

Here again an important condition for instability that the tooth passage frequency ($f_t = nz/60$) is slightly higher than the natural frequency of the workpiece. The expected band width of resonance is some 4–5% related to the resonance frequency of the workpiece.

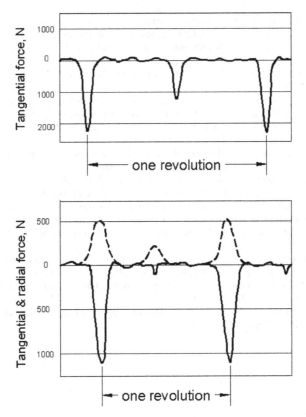

Fig. 5.46 Unequal tooth feed and cutting force due to workpiece vibration clamped on a CNC-router and using a tool with two cutting edges. Upper Figure: $n = 3000$ rpm, feed speed 5 m/min, cutting depth 1 mm. Bottom Figure: $n = 6000$ rpm, feed speed 5 m/min, cutting depth 1 mm (Csanády and Németh 2005)

Literature

Berolzheimer, C., Best, C.: Improvements through research on thin circular saw blades. Forest Prod. J. 404–412 (1959)

Csanády, E., Németh, S.: Investigation of clamping on CNC router. In: Proceedings of 17th IWMS Rosenheim, pp. 456–471 (2005)

Gogu, G.: Elastic stability and vibration on circular saws. In: Proceedings of 13th IWMS Vancouver, pp. 181–192 (1997)

Lehmann, B., Hutton, S.: The kinematics of wasboarding of bandsaws and circular saws. In: Proceedings of 13th IWMS Vancouver, pp. 205–216 (1997)

Lister, P., et al.: Experimental sawing performance results for super critical speed circular saws. In: Proceedings of 13th IWMS Vancouver, pp. 129–147 (1997)

McKenzie, W.: The effects of slots on critical rim temperature. Wood Sci. **4**, 304–311 (1973)

Mote, C.: Free vibration of initially stressed circular disk. Trans. ASME 258–264 (1965)

Mote, C., Szymani, R.: Principal developments in circular saw vibration. Holz Roh Werkst. S. 189–196, 219–225 (1977)

Okay, R., et al.: What is relationship between tooth passage frequency and natural frequency of the bandsaw when wasboarding induced. In: Proceedings of 12th IWMS Kyoto, pp. 267–380 (1995)

Orlowski, K., Wasielewski, R.: Washboarding during cutting on frame sawing machines. In: Proceedings of 15th IWMS Los Angeles, pp. 219–228 (2001)

Pahlitsch, G., Friebe, E.: Über das Vorspannen von Kreissägeblättern. Holz Roh Werkst. S. 429–436, S. 457–463 (1973)

Satoru, N.: Hyper critical sawblade. In: Proceedings of 16th IWMS Matsue, pp. 225–233 (2003)

Satoru, N.: Stable sawblade. In: Proceedings of 17th IWMS Rosenheim, pp. 418–420 (2005)

Sitkei, G., et al.: Theorie des Spanens von Holz. Fortschrittbericht No. 1. Acta Fac. Ligniensis Sopron (1990)

Stakhiev, Y.M.: Research on circular saw vibration in Russia: from theory and experiment to the needs of industry. Holz Roh Werkst. **56**(2), 131–137 (1998)

Ulsoy, A., Mote, C.: Analysis of bandsaw vibration. Wood Sci. **1**, 1–10 (1980)

Wu, W., Mote, C.: Analysis of vibration in bandsaw system. Forest Product J. 12–21 (1984)

Жодзишский, Г.А.: Влияние напражений от неравномерного нагрева, проковки и центробежных сил инерции на частоты свободных колебаний круглых пил (Influence of thermal stresses, tensioning and stresses from centrifugal forces on the natural frequency of circular saws). Ph.D. dissertation, Forestry Academy of Leningrad (1958)

Лапин, Л.: Определение допустимого числа оборотов круглых пил (Determination of allowable speed of circular saws). Лесной Журнал (2), с. 125–135 (1959)

Стахиев, Ю.: Работоспособностъ плоских круглых пил (Working capacity of circular saws) М.: Изд. Лесная Пром. 384 с (1989)

Стахиев, Ю.: Устойчивость и колебание плоских круглых пил (Stability and vibration of circular saws) Изд. Лесная Пром. (1977)

Chapter 6
The Stability of Wood Working Tools

6.1 Introduction

Processing can only form the expected shape, size and surface quality of a workpiece when the tool works stably and without excessive vibrations. The stability problems, (similar to the general engineering practice), occur in relation to thin and slender tools. These kinds of tools are primarily saws, band saws and circular saws with large diameters.

The main operational and tool design factors influencing cutting accuracy are summarized as follows:

- saw blade dimensions (width, thickness, diameters, free length of the cutting span, tooth pitch and shape),
- preparation of saw blades (roller-tensioning, heating, straining of a band saw, grinding accuracy),
- operational parameters (blade speed, tooth feed, cutting height, cutting depth, wood material properties, tooth sharpness).

Since the number of influencing factors are high, an optimum choice is not a simple task. On the other hand, the operator has any possibilities to improve sawing accuracy and to maximize feed speed.

The loss of the stability of a tool is indicated by the deviation from its original plane (Fig. 6.1). This blade deviation always causes skew cut and the width of the kerf also increases. This can lead in a certain case to the unacceptable decrease of the work-quality and cutting accuracy.

The deflection of the tool from its own plane is caused by eccentric and lateral forces on the cutting edge. The lateral forces especially are dangerous, because saws do not have much lateral stiffness.

The reasons for the development the eccentric and lateral forces on the cutting edge are the geometric irregularities of the succeeding cutting edges, the interaction between the blade's lateral surfaces and the workpiece, band wheel eccentricities and

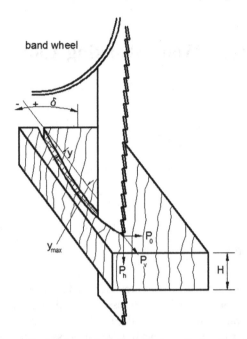

Fig. 6.1 The deflection of the band saw blade from the theoretical cutting plain

the inhomogeneity of the wood. The slender structural elements deform and buckle easily under compressive stresses. The tangential component of the heat stress developing in circular-saws is always compressive, and this stress component can cause waviness of the disk, the decrease of the natural frequency and the lateral stiffness (stability). The tangential compressive stress due to heat load can be decreased here with slotting along the periphery and with roll tensioning.

In this Chapter the main questions of machining accuracy, the stability problems of wood cutting tools, effect of heat load and tool preparation are discussed in detail.

6.2 Cutting Accuracy

The sawn surface always has irregularities due to several disturbing factors such as natural and forced vibrations, lateral forces, non-isotropy of wood, uneven tooth setting etc. These irregularities can be quite different. For example, washboarding (see in Sect. 5.4) is a very regular sinusoidal pattern that sometimes occurs on wood cut by band saws and circular saws. The depth of the corrugation varies from 0.2 to 0.6 mm. This dimensional variation requires the sawyer to cut the lumber thicker than normal to allow the planer to produce a smooth surface.

Uneven tooth setting and the non-isotropy of wood always cause an irregular surface which can be evaluated by statistical methods; (see Fig. 6.2). If the length

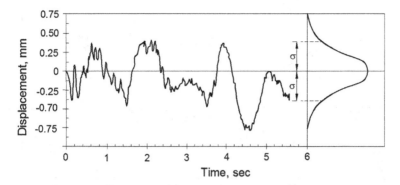

Fig. 6.2 Typical sawing variation profile measured from a sawn cant surface, its normal distribution and standard deviation

of the workpiece or time span is long enough for evaluation, then the collected data show a normal distribution and the standard deviation can be determined.

A slight inaccuracy in tooth angles can be detrimental to cutting performance of thinner saws. If the front face is out of square by only 0.5°–1.0°, the saw will increasingly lean to one side. The result is a skew cutting with considerable dimension errors. Manual swaging of teeth can often produce an uneven tooth setting with one side heavier than the other. This will also cause skew cutting.

Evaluating a set of measurements as a function of a given variable (e.g. feed speed), the standard deviations will be represented in a manner shown in Fig. 6.3 (Lister et al. 1997).

The points show always a scattering and the line which fits best gives the mean value. Here we can also define the standard deviation of the measured points and it is customary to use the "mean + standard deviation" value. As we see from the Figure, the cutting accuracy decreases at higher feed speeds due to higher forces on the teeth.

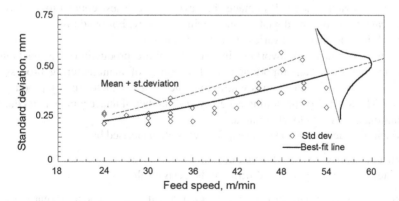

Fig. 6.3 Sawing accuracy versus feed speed for a circular saw 500 mm in diameter, blade thickness 2 mm, rotation speed 1900 rpm

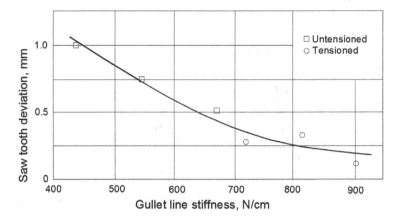

Fig. 6.4 Saw tooth deviation versus gullet line stiffness for a band saw with 1.5 m wheel diameter and 1.5 mm thick blade

One of the most important factors influencing sawing accuracy is the lateral stiffness related to the tooth tip or gullet line. Figure 6.4 shows such a relationship for a band saw with a wheel diameters of 1.5 m, and a band 260 mm wide and 1.5 mm band thick. The feed speed was 60 m/min and blade speed was 48 m/s and with cutting span length of 750 mm (Taylor and Hutton 1995). It is apparent from the Figure that a strong relationship exists between lateral edge stiffness and saw blade deviation and, hence, cutting accuracy.

The lateral edge stiffness is a complex parameter influenced by blade diameter or free span length, blade thickness and rolling stresses. The straining of the blade is also a very important factor in band saws.

In certain cases contacts may occur between the body of the bandsaw blade and the wood. The blade can move laterally in the kerf and if the side clearance (half the difference between the width of the tooth and the width of the blade) is smaller than some critical value, the blade contacts the sawn surface. These contact forces have a significant effect on saw deviation and sawing accuracy. For band mills, the critical value of the side clearance varies around 0.25 mm.

In the secondary woodworking industry (furniture production) the machining accuracy plays also an important role. The ability of manufacturing process to consistently provide dimensions within tolerance is a key factor in any production facility. The tolerance, machining accuracy and number of unacceptable parts due to machining errors are closely related.

Machining accuracy (machine capability) is characterised by

- stochastic error (standard deviation, σ),
- systematic error (typically shifting of the design value, $\pm\Delta\mu$).

Reasons for systematic errors typically include setting error and tool wear, causing a constant or continuously growing shifts in the process.

In order to be able to utilise machine capability, the machining process has to be made stable first by identifying and eliminating assignable causes of variation, then stability has to maintained by keeping the process under control. The tools of SPC (Statistical Process Control) serve these objectives.

The accuracy of machining depends on many influencing factors:

- machine rigidity and damping,
- accuracy of spindle running, bearings,
- clamping of workpiece,
- sharpness of tool,
- running circle accuracy,
- tooth bite,
- feed speed,
- relative workpiece mass (g/cm) in through-feed machines,
- mechanical properties of the processed material,
- spring constant of press rolls,
- resolution of setting mechanism.

These factors contribute to the stochastic error, except for the last one which manifests itself in systematic error. Variation of the machined dimensions with respect to the mean follows normal distribution when a number of causes are acting in a random way, each with small effect as related to the total variation. This is the case of the stochastic error and in this case the machining error is predictable. Using recent woodworking machines the standard deviation of stochastic errors can be kept within 0.05 and 0.1 mm (Csanády et al. 2019, and see also Sect. 9.5).

6.3 Blade Tensioning

The tensioning process used in the preparation of saw blades (band saws, circular saws and frame saw blades) has been in existence for about 100 years. The tensioning of blades involves the cold rolling of the blade in longitudinal or circumferential direction between two narrow rollers (Fig. 6.5).

The pressure produced by the rollers causes plastic deformation and permanent residual stresses in the blade material. The art of roller tensioning is to introduce these stresses in a manner that increases the stiffness and natural frequency of the blade and improves the cutting accuracy and feed speed of the saw. Different kinds of stresses may be present in a saw blade, such as initial stresses due to tensioning, rotational stresses and thermal stresses. These stresses lie in the saw blade plane and they are called *membrane stresses*. During deformation of the blade, these stresses do mechanical work and this work counteracts to a transverse displacement under a given external load. Therefore, an increase in the work of deformation of the stresses stiffens the saw blade. This contribution to the blade stiffness is very important with thin saw blades since the deformation work of the membrane stresses increases relative to the work done by the bending stiffness. (Mote 1965).

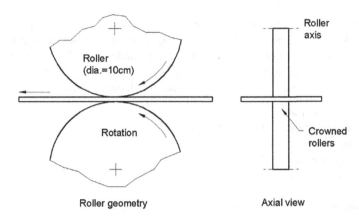

Fig. 6.5 Diagram of the roller tensioning process (Taylor and Hutton 1995)

It is well known that thin-walled structures are very sensitive to compression stresses causing buckling. Thermal stresses are especially dangerous because they dramatically decrease the stiffness of the tips of the teeth. These stresses are the dominant cause of saw instability. Increasing the rim temperature decreases the second and higher nodal diameter natural frequencies. As a consequence, the critical speed is reduced. Tensioning should counterbalance the compressive stresses. Therefore, tensioning is particularly significant for thin and large diameter saws where membrane stress effects on saw stiffness are large when compared to the bending stiffness.

Another important factor in band saws is the *blade position on the wheel*. Roll tensioning of the saw blade has been shown to significantly improve the precise positioning of the saw blade on the wheels. A stable blade is desirable in a high production band mill allowing high feed rates with minimal blade movement. Precise positioning of the blade due to tensioning does not require significant amounts of overhang, which have been shown to reduce the stiffness of the cutting edge of the blade. Without tensioning, too little overhang would have the risk of the blade moving back on the wheel and damaging the guides.

The roller crown is the fundamental part of any tensioning machine. During the rolling process, the rolling track should be compressed to establish tensile circumferential stresses around the rim in the tooth zone.

Depending on the shape (profile) of the roller, different $\varepsilon_t/\varepsilon_r$ strain ratios can be achieved, where ε_t is the circumferential strain along the rolling path and ε_r is the radial strain perpendicular to the rolling path. This strain ratio depends on the crown radius or on the width of the roller and on the depth of the rolling track. Their relationship is shown in Fig. 6.6 (Stakhiev 2001). In all cases the strain ratio $\varepsilon_t/\varepsilon_r$ decreases with increasing rolling depth.

The rollers generally have diameters around 70–100 mm and a crown radius of 50–100 mm. If a cylindrical surface is used, its width varies between 4 and 6 mm. The rolling position is characterised by the relative rolling radius (r_m/R) for circular

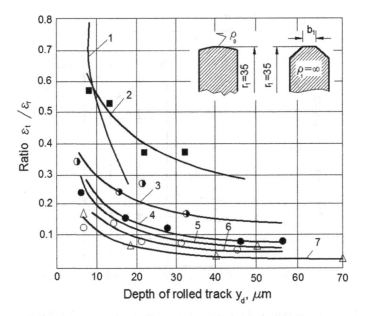

Fig. 6.6 Relationship between the depth of rolled track y_d and the ratio $\varepsilon_t/\varepsilon_r$: **1**—$\rho_1 = 350$ mm; **2**—$\rho_1 = \infty$, $b_1 = 10$ mm; **3**—$\rho_1 = 210$ mm; **4**—$\rho_1 = 105$ mm; **5**—$\rho_1 = \infty$, $b_1 = 6$ mm; **6**—$\rho_1 = 35$ mm; **7**—$\rho_1 = \infty$, $b_1 = 4.5$ mm

saws and by the symmetric distance from the centre line for band- and frame saws (Stakhiev 1966; Pahlitsch and Friebe 1973; Bajkowski 1967).

In circular saws, the stresses induced by rolling can be calculated by the following equations:

for the inner part

$$\sigma_{ri} = -\frac{\delta \cdot E}{2 \cdot r_m} \frac{R^2 - r_m^2}{R^2 - r_c^2}\left(1 - \frac{r_c^2}{r^2}\right) \tag{6.1}$$

$$\sigma_{ti} = -\frac{\delta \cdot E}{2 \cdot r_m} \frac{R^2 - r_m^2}{R^2 - r_c^2}\left(1 + \frac{r_c^2}{r^2}\right) \tag{6.1a}$$

for the outer part

$$\sigma_{ro} = \frac{\delta \cdot E}{2 \cdot r_m} \frac{r_m^2 - r_c^2}{R^2 - r_c^2}\left(1 - \frac{R^2}{r^2}\right) \tag{6.2}$$

$$\sigma_{to} = \frac{\delta \cdot E}{2 \cdot r_m} \frac{r_m^2 - r_c^2}{R^2 - r_c^2}\left(1 + \frac{R^2}{r^2}\right) \tag{6.2a}$$

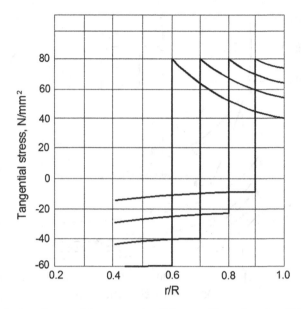

Fig. 6.7 Distribution of tangential stresses using different rolling ratios. $D = 500$ mm, clamping ratio 0.3 and rolling force 16 kN

where R is radius of the blade, r_m is the rolling radius and r_c is the clamping radius. δ is the radial strain due to rolling and its approximate value is calculated as

$$\delta = 3.0 \frac{F_r^2}{s^2}\left(0.6\frac{r_m}{R} - 0.1\right)$$ (6.3)

where the rolling force F_r in kN and the plate thickness s in mm must be substituted.

The typical stress distributions for a circular saw 500 mm in diameter are seen in Fig. 6.7 (Pachlitsch and Friebe 1973). Tensile stresses occur outside the rolling radius, which counteract thermal compressive stresses, stabilize the blade and increase blade stiffness. However, the rim tensile stress slightly decreases with decreasing rolling radius. With an increasing strain ratio $\varepsilon_t/\varepsilon_r$, the stress distribution around the rolling radius will slightly be modified.

There is a maximum rolling force, at which just no static buckling occurs. This load is the *critical rolling force*. The critical rolling force depends on the rolling radius ratio and its value for a given saw diameter and thickness is shown in Fig. 6.8 (Stakhiev 2001). In practice, rolling forces slightly less than the critical values are generally used.

Tensioning stresses increase the second and higher nodal diameters natural frequencies. This favourable effect of tensioning stresses depends on the rolling radius. Generally, optimum results can be achieved between rolling ratios of 0.7 and 0.85, as it is shown in Fig. 6.9 (Stakhiev 2001). If constant rolling force is used, then the optimum rolling radius is slightly shifted towards smaller rolling radii.

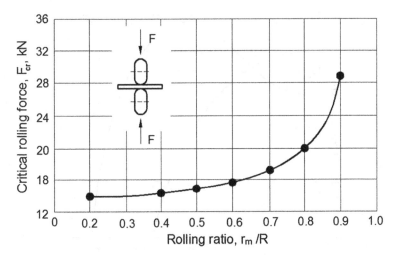

Fig. 6.8 Influence of rolling ratio r_m/R on the critical rolling force F_{cr} corresponding to the mode with zero nodal diameter ($n = 0$) and zero nodal circles ($m = 0$); $D = 500$ mm, $h = 2.2$ mm, $z = 48$

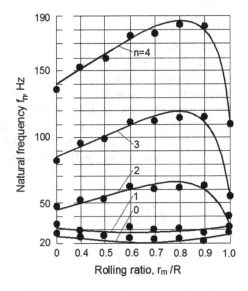

Fig. 6.9 Determination of the optimum ratio of rolling for rolling along one track: saw diameter = 700 mm, thickness = 3.2 mm, clamping diameter = 160 mm, critical rolling force for each track

It should be noted that the tensioning stress always decreases the static blade stiffness so the favourable effect of tensioning can be utilized with higher rotation speeds. Figure 6.10 shows the lateral stiffness of a saw blade as a function of rotation speed with and without tensioning (Bajkowski 1967; Stakhiev 2000).

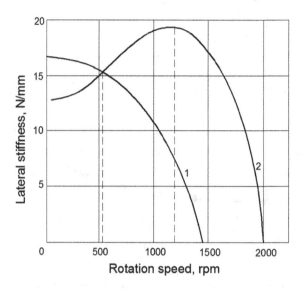

Fig. 6.10 Saw blade stiffness as a function of rotation speed. ($D = 800$ mm, $s = 2.8$ mm) **1**—without tensioning, **2**—critically tensioned

The stiffness of the untensioned blade gradually decreases and tends to zero at the critical speed. The tensioned blade first has an increasing stiffness but after reaching its maximum value, it decreases and tends to zero. The two curves intersect and this rotation speed is sometimes called "the universal rotation speed." The critical speed increment due to tensioning is generally between 20 and 30%.

Another method to induce membrane stresses is to use internal pressure acting on the internal circumference of the disk. This has an effect equal to the traditional roller tensioning method. The in-plane stresses will be obtained by the following equations:

$$\sigma_r = -\frac{r_i^2}{R^2 - r_i^2}\left(1 - \frac{R^2}{r^2}\right) \cdot p_i \tag{6.4}$$

$$\sigma_t = \frac{r_i^2}{R^2 - r_i^2}\left(1 + \frac{R^2}{r^2}\right) \cdot p_i \tag{6.4a}$$

where r_i means the internal radius of the disc and p_i is the internal pressure. The induced pressure distribution is similar to those of roll tensioning outside the rolling path (Fig. 6.11) (Sanyev 1980).

Centripetal tensioning is another tensioning possibility (Renshaw 1999). Centripetal tensioning introduces in-plane residual stresses into the saw blade only while the saw is rotating. This is achieved by a clamp with pivoting wedges that rest against the inner edge of the saw blade, Fig. 6.12 (Renshaw 1999). The advantage of centripetal tensioning is that the residual stresses are only present when there are

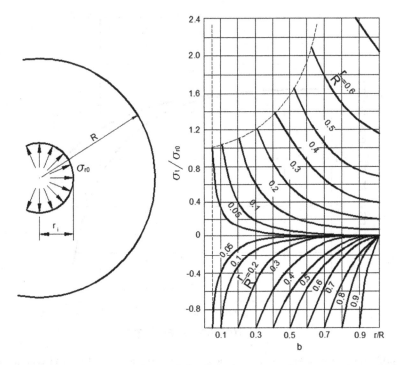

Fig. 6.11   Distribution of tangential and radial stresses due to internal pressure acting on the inner circumference of a circular saw

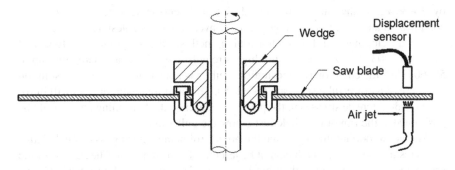

Fig. 6.12   A schematic of a centripetal tensioning clamp and the experimental set up used to measure transverse stiffness

also centripetal stresses due to rotation. Furthermore, at increasing rotation speeds the centripetal tensioning stresses will also be increased. Theoretical calculations and experimental measurements have verified the effectiveness of centripetal tensioning.

Figure 6.13 shows the lateral stiffness of a circular saw with and without centripetal tensioning. Using this tensioning method, the critical rotation speed can be increased

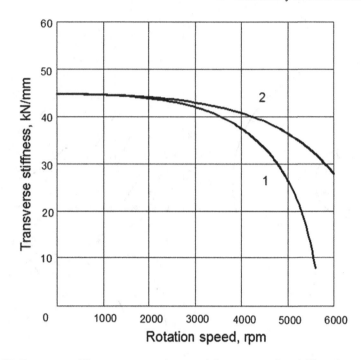

Fig. 6.13 Transverse stiffness versus rotation speed for an untensioned (**1**) and centripetally tensioned (**2**) saw. (Diameter 30.5 cm, blade thickness 1.2 mm, clamping ratio 0.44)

by 20–30%. At the same time, the design and tensioning mechanism is more complicated and expensive compared to a conventional blade design.

The development of further tensioning methods are in progress (Münz and Thiessen 2003; Münz 2005). Using high strength steels with case hardening up to 58–60 HRC, the common roll tensioning method is not suitable. A possible solution is the use of a powerful laser beam (5–6 kW). The laser treatment of steel will introduce residual tensile stresses in the path of the laser so the tensioning path should be positioned at the rim of the blade just below the gullet line.

Another important finding is that the process of manufacturing a saw blade introduces residual stresses, which cannot be left out of consideration. The surface of the saw blade can be ground with cubic boron nitride (CBN) grinding wheels (Münz 2005; Matalin 1954). Grinding of the blade surface causes residual stresses in the near surface layer. They may vary from compressive to tensile values in relation to the grinding depth. The presence of compressive residual stresses in the near surface layer has a damping effect, resulting in smaller vibration amplitudes.

Band saws are also subjected to tensioning to improve blade stability and blade positioning on the wheel. The resultant stress in a band saw is composed of various stress components. The main components are shown in Fig. 6.14 with their distribution along the band width. The conventional roll tensioning occurs along the back line and the gullet line within 2 cm of the rim. Tension stresses occur outside the

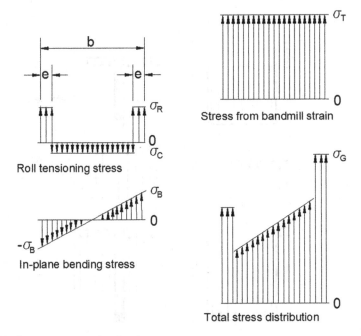

Fig. 6.14 Stress components in a bandsaw

rolling path while in the central part compression stresses dominate. The induced tensile stresses are around 60 N/mm².

The in-plane bending stresses result from wheel tilt, the wheel crown and blade overhang. The third main component is band mill strain having values from 70 to 100 N/mm² or even higher.

The degree of tensioning is determined by measuring the deformations of the blade (transverse curvature). In addition, a longitudinal curvature in the plane of the blade may be introduced by rolling near to the back edge. This in-plane curvature is called *backcrown* and its equivalent radius is 700–800 m (Taylor and Hutton 1995).

A modified roller tensioning procedure coupled with an increase in band mill strain was also proposed to improve blade stiffness and sawing accuracy over convention-ally tensioned blades (Taylor and Hutton 1995). The basic idea of this proposal is that keeping the rolled region away from the cutting edge of the bandsaw achieves the most effective tensioning (Bajkowski 1967). Therefore, a centre tensioning method is proposed. The two tensioning methods are compared in Fig. 6.15 (Taylor and Hutton 1995). Due to the higher distance between the rolling path and the gullet line, the tensile stress induced in the cutting region will be smaller requiring a higher band mill strain. The overall stress level is limited by fatigue life considerations, such as gullet cracking, and this will set an upper limit on the magnitude of the combined stress. The maximum combined stress is generally limited to 150 N/mm².

A more detailed rolling profile for centre line tensioning is given in Fig. 6.16 (Taylor and Hutton 1995). Rolling occurs in a narrow centre region 40 mm wide. 5

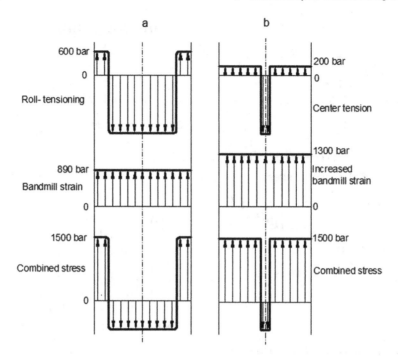

Fig. 6.15 Diagram of stresses for conventional and centre tensioned saws, **a** conventional tensioning, **b** centre tensioning

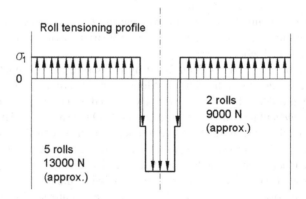

Fig. 6.16 Rolling profile for centre tensioned blades

rolling passes in the centre line region are made with 14 kN rolling force and final pair of rolling passes is made at a reduced rolling force of 9 kN, softening the transition from the rolled region to the unrolled region. Centre line rolling ensures a sufficient transverse curvature so that the saw blade is stable on the band mill wheels. The maximum displacement in the centre of saw blade due to the transverse curvature should be around 0.5 mm for a band 260 mm wide and a blade 1.5 mm thick. This value is considered adequate to keep the blade stable on the wheels.

In a running machine, the change of axial band stress due to centripetal acceleration may occur depending on the straining mechanism of the upper wheel.

The centripetal force acting on the blade depends on the blade speed and wheel radius

$$F_c = m \cdot \frac{v^2}{R}$$

Integrating the elementary forces along the semi-circular arc we get

$$F_c = \rho \cdot 2b \cdot s \cdot v^2 \tag{6.5}$$

where ρ is the density, b is the width and s is the thickness of the blade.

This force related to the blade cross section yields the centripetal stress

$$\sigma_c = \rho \cdot v^2$$

The centripetal stress decreases the straining stress if its effect is not compensated for by using an appropriate straining mechanism.

The actual stress of a saw blade must be known to determine the fatigue life of a saw blade. The stress loading in the blade varies considerably during one blade revolution due to bending stresses. A blade running on a wheel is subjected to bending and the bending stress can be calculated as

$$\sigma_M = \varepsilon \cdot E = E \frac{s}{D} \tag{6.6}$$

where D means the wheel diameter.

A typical loading cycle for one blade revolution is shown in Fig. 6.17. The loading of the blade is highly cyclic with large amplitude often causing gullet cracking.

The stress in the cutting span is always higher than that of the non-cutting span. These stresses are balanced by the straining and centripetal stresses:

$$\sigma_1 + \sigma_2 = 2 \cdot (\sigma_s + \sigma_c)$$

where σ_s is the straining stress in the blade. The Figure shows the maximum stress in the blade which is

$$\sigma_{\text{max}} = \sigma_2 + \sigma_M$$

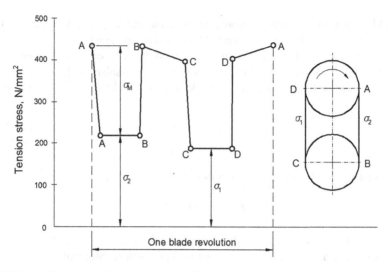

Fig. 6.17 Loading in a band saw blade for one blade revolution

6.4 The Stability of the Band Saw

The forces acting on a single cutting edge of a band saw are given in Fig. 6.1. The P_h cutting force acts in the plane of the blade and along its length, so it can be neglected, as a deflection force.

The P_v feed force may be central or eccentric. The central force does not turn the blade from its plane, but the eccentric feed force is already a deflection force. The most important effect is the lateral force which is also a deflection force.

We have summarized the deformations (deflections), caused by the forces acting on the edge in Fig. 6.18 (Pahlitsch and Putkammer 1974). The deformation caused by an eccentric force can be put together from two parts. Placing the force into the centre line we get a parallel displacement, to that we add the rotation caused by the $M = P_0 \cdot a$ bending moment (see Fig. 6.18b, c).

The deflection caused by the lateral force can be calculated as follows: The parallel displacement of the blade Fig. 6.18 (Pahlitsch and Putkammer 1974):

$$x = \frac{P_0}{4 P_e}\left(L_e - 4\sqrt{\frac{R}{P_e}}\right) \quad \text{and} \quad R = \frac{E \cdot I}{1 - \nu}$$

the stiffness of the lateral displacement is given by:

$$c_x = \frac{P_0}{x} = \frac{4 P_e}{L_e - 4\sqrt{\frac{R}{P_e}}} \tag{6.7}$$

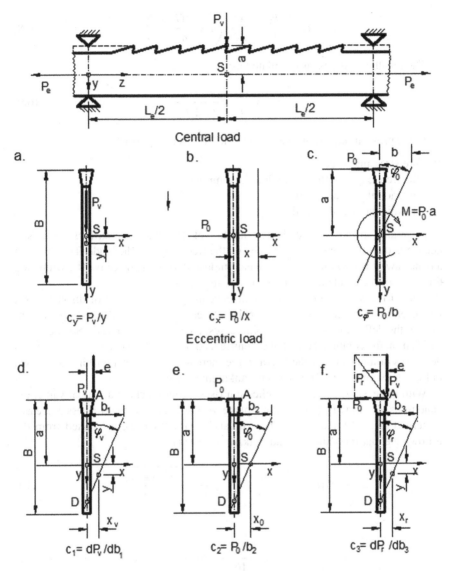

Fig. 6.18 Different possible forces acting on the edge of a band saw and the definition of appropriate stiffness coefficients

The rotation of a saw blade around the centre of gravity (Fig. 6.18):

$$b = a \cdot \varphi_0 = \frac{0.75 \cdot P_0 \cdot L_e \cdot a}{a \cdot P_e + 4G \cdot s^3}$$

and the stiffness of rotation:

$$c_\varphi = \frac{P_0}{b} = \frac{4P_e}{3L_e}\left(1 + \frac{4G \cdot s^3}{a \cdot P_e}\right) \tag{6.8}$$

The resultant stiffness is calculated as:

$$c_2 = \frac{P_0}{b} = \frac{c_x \cdot c_\varphi}{c_x + c_\varphi} \tag{6.9}$$

In the previous equations (see also the notations in Fig. 6.18).

P_e pre-stressing force,
L_e the free length of the band without clamping,
G the modulus of elasticity in shear,
s the thickness of the band.

A greater stiffness causes better stability of the band. The stiffness is mostly determined by the pre-stressing force and the free length of the blade. Therefore on a band saw, it is important to use a band guide and to set it correctly. The width and thickness of the band has less influence on its stability.

The stiffness has remained constant depending as a function of displacement. However, eccentric feed forces (Fig. 6.18d, f), cause the most stiffness at the beginning of the deflection. As deflection increases, the stiffness decreases. Stiffness also varies from the combined effect of eccentric feed force and the lateral force. Stiffness decreases dramatically as the lateral force increases. Their relationships are shown in Figs. 6.19 and 6.20 (Pahlitsch and Putkammer 1974).

With increasing feed forces, deflection rises. After reaching a critical value, the band turns 90° away from its original plane. At the same time the stiffness decreases to zero (Figs. 6.19 and 6.20). The critical feed force can be calculated from the following equation (Pahlitsch and Putkammer 1974):

$$P_{vcr} = -\frac{a \cdot c_x}{2k^2} + \sqrt{\frac{a^2 \cdot c_x^2}{4k^2} + \frac{a^2 \cdot c_x \cdot c_\varphi}{k^2}} \tag{6.10}$$

where:

$$k = \frac{4 + \pi^2}{16} = 0.867$$

Band saw blades are subjected to roller tensioning to improve blade stability and blade positioning on the wheel. Roller tensioning stresses (Figs. 6.14 and 6.15) considerably increase the lateral blade stiffness and decrease the torsional vibration amplitudes.

Since there is a strong relationship between the lateral edge stiffness and cutting accuracy, the resultant stress distribution in the blade should provide the highest possible lateral edge stiffness (see Fig. 6.4).

In practical cases, the feed speed is an important factor determining sawing performance. An increasing feed speed always increases the feed force (radial cutting

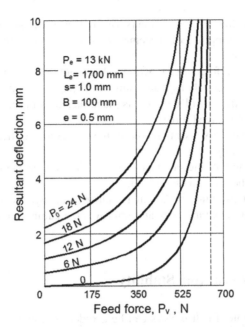

Fig. 6.19 The resultant deflection of a band saw depending on the feed force for different lateral forces

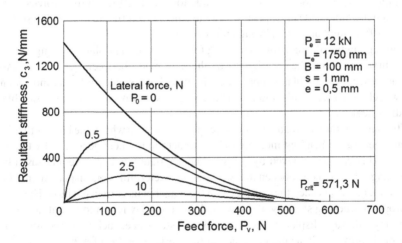

Fig. 6.20 The resultant stiffness of a band saw depending on the feed force

force) and its lateral components causing higher tooth tip deflection. Both tooth tip deflection and sawing accuracy are strongly related to the feed speed (see Fig. 6.3).

Based on theoretical considerations and practical observations, the following main conclusions may be drawn:

- a precise and careful tool preparation (spring or swage setting, grinding accuracy) highly decreases the presence of eccentric and lateral forces, which would increase tooth tip deflection;
- eccentric and lateral forces due to wood failures and inhomogeneity of the timber can be decreased primarily by decreasing of the feed per tooth;
- the deflection of the saw blade from its plane increases linearly under the effect of the increase of the feed force;
- the increase of the pre-stressing increases the value of the critical feed force, and improves the stability;
- the free span length of the saw blade must be as small as possible, as the band guide prevents the rotation of the blade;
- introduction of tensioning stresses in a manner that increases the lateral stiffness of the blade in a favourable way and improves the cutting accuracy of the band saw.

6.5   Band Saw Tracking Stability

The sideways motion of belt pulleys can be avoided by using crowned wheels. A saw blade running on the band mill wheels is subjected to feed forces, which tend to move the saw blade "front-to-back." This sideways motion of the saw blade is called *tracking*. The ability of a moving saw blade to counteract the cutting forces and to maintain its initial and stable position on the wheel is referred to as *tracking stability* (Swift 1932; Sugihara 1977; Wong and Schajer 1997).

During sawing, the feed forces (cutting forces) create an in-plane bending moment and curvature along the free length of the saw blade between the wheels. As a consequence, the saw blade approaches the wheel at a small angle and moves onto the wheel with a screw-like motion. This causes a sideways motion of the saw blade on the wheel.

To maintain the saw blade in its correct position on the wheel, the band mill wheel should generate a bending moment to counteract the moment generated by the feed forces. This can be achieved by using an appropriate wheel profile and saw blade overhang. The balance between the two moments determines the displacement of the saw blade and it can keep the saw blade on the wheel in its correct position. Having an appropriate tracking stability, the saw blade will quickly return to its initial position after any sideways displacement caused by a disturbance, such as a feed force. The saw blade moves only a small distance in response to the feed force.

Figure 6.21 shows a crowned band mill wheel and its circle profile. The profile can easily be described in the following manner:

$$f(y) = h_c \cdot \left(1 - \left(\frac{y - \frac{b}{2}}{\frac{b}{2}} \right)^2 \right)$$

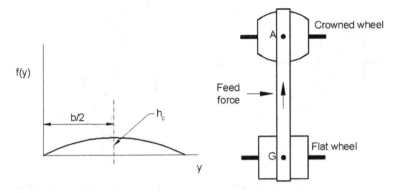

Fig. 6.21 Crowned band mill wheel and its profile (Wong and Schajer 1997)

where b is the width of the wheel and h_c is the crown height. The crown height is generally two per thousand of the width of the wheel.

There are different ways to achieve appropriate band saw stability. Figure 6.22 shows several possibilities, such as a crowned wheel, a tapered wheel and band overhang. Tilting a flat wheel corresponds to a tapered wheel.

The feed force (cutting force) generates a bending moment in the band. Therefore, the band between the wheels behaves as a beam bending under an axial tensile load with a centre force acting in the span. The bending moment at any point along the length of the band is

$$M_1 = E \cdot I \frac{d^2 y}{dx^2}$$

where $E \cdot I$ is the flexural rigidity of the band and $\frac{d^2 y}{dx^2}$ gives the band curvature.

The mechanism of band saw tracking stability can be examined on a tapered wheel shown in Fig. 6.23 (Wong and Schajer 1997). The tapered wheel applies the

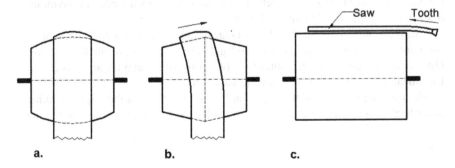

Fig. 6.22 Crowned wheel (**a**), double tapered wheel (**b**) and the tracking effect of overhang (**c**) (Wong and Schajer 1997)

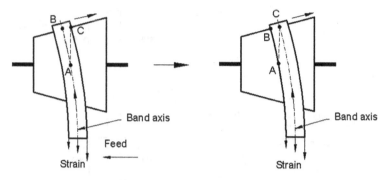

Fig. 6.23 Band motion on a tapered wheel, initial position and position after 1/4 turn (Wong and Schajer 1997)

strain unevenly to the band, which results in a bending moment and consequently a curvature in the band. The first contact point **A** on the wheel tends to follow a path perpendicular to the rotational axis to point **C**. Therefore, the band moves sideways towards the wider side of the tapered wheel.

To understand the behaviour of a crowned wheel, it may be replaced by a double tapered wheel; (see Fig. 6.22b). Since the band is running on the left tapered side, this side tends to move the band towards the centre line. As the band approaches the centre of the wheel, the influences of the left and right taper sides will be equalized and the band will reach an equilibrium position. Similarly, the band running on a crowned wheel tends to remain at the top of the crown. That means, the crowned wheel centres itself. In Fig. 6.22c, we see a band overhanging the flat wheel that curls down towards the wheel axis. The curling of the band has a similar effect as that of the tapered wheel and creates also an uneven stress distribution, and a bending moment in the band. Therefore, the use of a band overhang on a flat wheel will also centre itself.

To counterbalance the bending moment of the feed force, the band should be running slightly asymmetrically on the crowned wheel, ensuring the appropriate bending moment in opposite direction. With increasing feed force, the asymmetry of the band on the wheel increases linearly.

Tilting the wheel sets the band overhang. In the vicinity of a zero tilt angle, the overhang is small and the position of the band is very sensitive to small tilt changes. Over a 0.1° tilt angle, the relationship between overhang and tilt angle is more or less linear; the overhang increases with wheel tilt.

The tracking stability of the band can be increased by combining the overhang and wheel crown. The band will not move sideways as much in response to a feed force.

6.6   Stability of Circular Saws

Saws operating in different conditions always vibrate, but this does not neces-
sarily mean instability. A sign for impending transverse instability is when any
small disturbance causes large transverse displacements of the saw from its current
equilibrium.

Two different instability mechanisms have been observed in saw vibration. The
first mechanism is the static buckling, where the saw blade has a constant harmonic
shape with large amplitudes. This mode of buckling consists of low nodal diameters
(from 0 to 1 nodal diameter) and the blade often rubs against the workpiece. The latter
action is associated with intensive frictional heating which causes more instability
of the blade.

The second instability mode is the critical speed instability (see in Sect. 5.3),
where the backward wave frequency equals the rotational speed frequency causing
a standing wave; (see Fig. 5.9). The critical rotation speed determines the maximum
stable rotation speed of the saw which in the practice is 15% less than the critical
speed.

The critical speed instability always occurs before buckling and is the more
significant and important instability mode.

Saw vibration has many drawbacks: increasing the kerf loss and energy requirements, reducing cutting and product accuracy, diminishing surface quality etc. Therefore, efforts should be made to reduce vibration amplitudes to a low level. This can
be accomplished through appropriate saw design and membrane stress modification.

The vibration amplitudes strongly depend on the lateral stiffness of the saw blade.
The static lateral stiffness can be calculated in the following manner (Sanyev 1980):

$$I_{st} = \frac{F}{y_{st}} = \frac{E \cdot s^3 \sqrt{R^2 - r_c^2}}{25(R - r_c)^3} \text{ N/m} \tag{6.11}$$

where E is the modulus of elasticity, N/cm^2, s is the blade thickness, cm, R is the
outer radius of the blade, cm, and r_c is the clamping radius, cm.

The above equation shows that the blade thickness has a very dominant influence
on its lateral stiffness. In big diameters saws, it is important to introduce appropriate
membrane stresses to increase their natural frequency and lateral stiffness; (see in
Sect. 5.2).

As shown in Fig. 5.11, different membrane stresses occur in a rotating disk
influencing its vibration mode, critical speed and also the stability of the saw blade.

The rotation of the saw induces only tensile stresses, which increase the stability of
the saw. This can also be seen in Fig. 5.9, where natural frequency of the saw increases
monotonically with its rotation speed. However, if the blade has unbalanced masses,
then this moving load can cause instability when there is moving load resonance at
critical speed.

A more important factor is the presence of thermal stresses. The magnitude of
thermal stresses depends on the coefficient of thermal expansion of the blade material

and the temperature distribution in the blade induced by friction forces. The thermal stresses in a tangential direction are always compressive and they are the dominant cause of saw instability. An increasing rim temperature decreases the natural frequencies of saw blades with nodal diameters of two or more, consequently reducing the critical speed. The saw becomes less stable at lower rotation speeds. Different measures can be used to counteract the unfavourable effect of thermal stresses. One way is to use an induction heater to heat the central region of the blade and produce thermal tensioning. This modifies the temperature distribution of the blade and lowers the temperature difference between the central and outer regions of the blade. Thermal stress depends on the temperature difference but not on the temperature itself.

Another frequently used measure is the use of slots along the periphery of the blade. The slots reduce compressive hoop stresses, allowing free expansion of the periphery without developing tangential stresses. The slots have a very complicated effect on blade stiffness, natural frequencies and vibration amplitudes. Therefore, contradictory experimental results were often obtained with slotted saws.

An interesting phenomenon associated with slots is that the number of resonant frequencies is doubled because of the frequency splitting due to the slots. At the same time, the excitation energy is more distributed throughout the spectrum, reducing the vibration amplitude compared to a saw without slots.

Tensioning, as discussed in detail in Sect. 6.3, is an artificially introduced residual stress in the saw through plastic deformation by rolling and/or through local heating. Optimum tensioning means the introduction of a favourable stress state, increasing the natural frequencies and the critical speed of the saw blade.

The main practical problem is that we do not have accurate non-destructive measurement methods to determine initial stresses in a saw blade. Therefore, two indirect methods of residual stress evaluation are generally used. Either we measure the natural frequencies of the saw and critical speed or its stiffness. The results of both measurement methods generally correlate to one another, so stiffness can be regarded as a measure of tensioning effect on the natural frequencies of a saw and its critical speed. The lateral stiffness of a saw can be measured under static and dynamic conditions. A dynamic measuring method is needed to obtain the stiffness characteristics as a function of rotation speed (see Fig. 6.12).

Thermally induced stresses are highly dependent on the operational parameters, such as the cutting height, feed speed and rotation speed. Therefore, the full compensation for these stresses by a constant tensioning stress is not possible. Attempts have been made to overcome this problem through continuous adjustment of the state of stress using on-line saw stability control (Sanyev 1980). Temperature differences along the blade radius can be eliminated by either local heating or cooling (Fig. 6.24).

Using temperature or displacement sensors, the duration or intensity of heating/cooling can easily be regulated. The displacement sensor measures vibration amplitudes, which indirectly refer to increasing temperature differences. Figure 6.25 shows the temperature course of a saw blade without and with water cooling during sawing. Appropriate cooling can practically eliminate temperature differences in a saw blade.

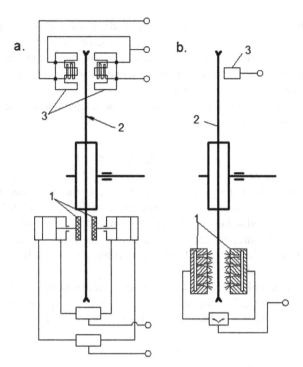

Fig. 6.24 On-line stability control of a saw blade using heating (**a**) or cooling (**b**). **a** heating with rubbing pads, **b** cooling with water jets (Sanyev 1980)

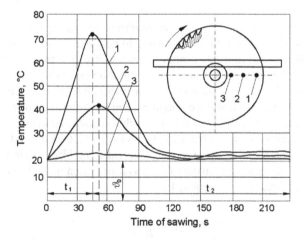

Fig. 6.25 Temperature course of the blade without cooling (t_1) and with continuous water cooling (t_2)

The results in Fig. 6.25 make it possible to draw some conclusions regarding thermal inertia of saw blades. In the transition from one operating condition to another the saw blade exhibits a thermal inertia, which depends on the geometrical dimensions of the blade and on its temperature. That means, under varying operational conditions warming-up and cooling-down periods occur.

After a sudden load increase (step function), the temperature in the blade varies with the following equation:

$$\vartheta = \vartheta_0 + (\Delta\vartheta)_{\max}\left(1 - e^{-\frac{t}{T}}\right) \tag{6.12}$$

where ϑ_0 is the ambient temperature and $(\Delta\vartheta)_{\max}$ is the maximum temperature rise until a steady-state condition. The time constant T characterizes the speed of temperature change following a load change. More exactly, the value of T indicates how much time is required to reach 63.5 per cent of the maximum temperature rise $(1 - 1/e)$. This value is $T \approx 30$ s slightly increasing towards the inner parts due to the decreasing temperature.

Generally we should not reckon with a creep relaxation of the original stress state due to roller-tensioning. That means that a saw should be rolled only once and this initial stress state with cyclic loading remains the same over time and it shows no tension decay.

6.7   Frame Saw Stability

Frame sawing machines in many European countries rank among the basic equipment for primary wood processing. It is interesting to note that the first ancient saw mill used in the fifteenth and sixteenth centuries was a man-driven frame saw, which was first depicted in Bessoni's "Theatrum Instrumentorum et Machinarum" in 1578.

Similarly to other sawing machines, the stiffness and stability of the saw blade determine sawing accuracy, surface roughness and the occurrence of washboarding. The saw blade can be regarded as a beam with a rectangular cross section with a free length L, blade width b and blade thickness s (Fig. 6.26). Both ends of the blade are clamped and strained with a force F. Generally the blade is strained with an eccentricity e creating a more favourable stress distribution in the blade.

The straining stress σ_0 is calculated as

$$\sigma_0 = \frac{F}{b \cdot s} \tag{6.13}$$

which generally amounts to between 100 and 200 N/mm^2.

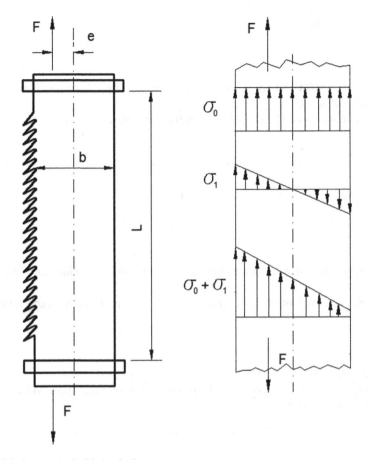

Fig. 6.26 Stresses in the blade of a frame saw

Eccentric loading generates a pure bending moment and, as a consequence, a bending stress

$$\sigma_1 = \frac{F \cdot e}{I} \cdot \frac{b}{2} \quad \text{with} \quad I = \frac{s \cdot b^3}{12}$$

and

$$\sigma_1 = \frac{6F \cdot e}{s \cdot b^2} \tag{6.14}$$

The resultant stress at the gullet line is

$$\sigma_g = \sigma_0 + \sigma_1$$

while along the back line it amounts to

$$\sigma_b = \sigma_0 - \sigma_1$$

This latter resultant stress may not be negative, which would mean a compressive stress causing instability.

The neutral axis will be shifted to the right side with a value of

$$n_a = \frac{b}{2}\frac{\sigma_0}{\sigma_1} = \frac{b^2}{12e} \tag{6.15}$$

If $e = b/6$, then the above equation yields

$$n_a = \frac{b}{2}$$

which means that the neutral axis lies on the back rim of the blade having a zero resultant stress.

The elongation of the blade under strain forces amounts to a measurable quantity:

$$\Delta L = \frac{\sigma_0}{E}L$$

which can be used to control the expected strain force.

During woodworking operations the blade temperature increases, which causes elongation of the blade

$$\Delta L_t = \alpha \cdot L \cdot \Delta\vartheta$$

where α is the thermal expansion coefficient (for steel $\alpha = 12 \times 10^{-6}$), and $\Delta\vartheta$ is the temperature difference. The elongation of a saw blade decreases the initial strain stress σ_0 with the following value

$$\Delta\sigma = -\alpha \cdot E \cdot \Delta\vartheta$$

For example, setting a strain stress $\sigma_0 = 150$ N/mm^2 and the temperature rise is $\Delta\vartheta = 60\,°C$, then $\Delta\sigma \approx 150$ N/mm^2 entirely stopping the strain on the blade.

The most common vibration mode of a saw blade is lateral although torsional vibration can also occur. The possible lateral vibration modes are shown in Fig. 6.27.

Due to the clamping of the blade at both ends, the end points are always nodal points. The natural frequencies can be calculated from the following equation:

$$f_n = \frac{n}{2L}\sqrt{\frac{n^2\pi^2}{L^2}\frac{E \cdot I}{\rho \cdot b \cdot s} + \frac{F}{\rho \cdot b \cdot s}} \tag{6.16}$$

where ρ is the density of the blade material and $n = 1, 2, 3$ denotes the number of vibration modes. The flexural rigidity of the blade $E \cdot I$ is given as

$$E \cdot I = E \frac{b \cdot s^3}{12}$$

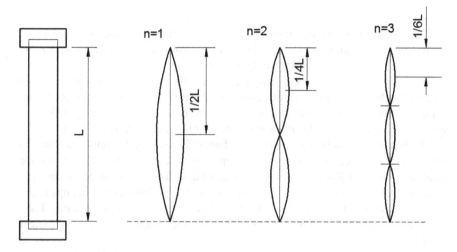

Fig. 6.27 The first three vibration modes of a blade clamped on both ends

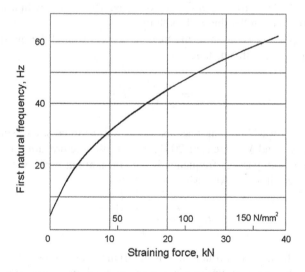

Fig. 6.28 First natural frequency versus strain force relationship. $L = 1.2$ m, $b = 115$ mm, $s = 2$ mm

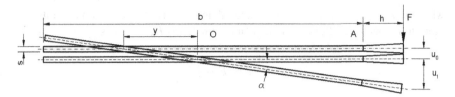

Fig. 6.29 Tooth tip deflection of a saw blade subjected to lateral load

Figure 6.28 shows the first natural frequencies for a given blade as a function of strain force (or strain stress). The relationship is not linear and at zero strain, the blade has a low natural frequency, only 3.3 Hz.

The main component of the cutting force lies in the saw blade plane and it causes no lateral deflection. However, the cutting force always has smaller or larger lateral component resulting in lateral displacement of the tooth tip. The tooth tip deflection will be determined by the lateral stiffness of the blade and its lateral forces.

The behaviour of a frame saw blade to lateral forces is quite similar to that of a band saw blade; (see in Sect. 6.4). The approximate calculation method given for band saw blade can also be used for frame saw blade. The behaviour of a frame saw blade subjected to lateral forces is given in Fig. 6.29. The deflection of the tips of the saw teeth has two components due to the blade lateral motion and blade rotation. Each component, u_0 and u_1, can be calculated and also the resultant stiffness coefficient can be determined.

The strain force has a fundamental influence on both the tooth tip deflection and stiffness. This relationship for a given blade is given in Fig. 6.30. The tooth tip deflection is calculated for a lateral force of 10 N. The tooth height h is defined as the distance between gullet line and tooth tip.

The actual point of rotation of the blade is always behind the mid-point of the blade (see Fig. 6.29) with a distance of

$$y = \frac{u_0}{u_1}\left(\frac{b}{2} + h\right) \tag{6.17}$$

In some cases washboarding was also observed in frame saws using mini saw blades (Orlowski and Wasielewski 2001). Due to the reciprocating motion of the frame saw, the tooth passage frequency is not constant but varies within one cycle according to the following expression

$$f_t = \frac{\pi \cdot f_r \cdot H \cdot \sin\varphi}{p} \tag{6.18}$$

where f_r is the frequency of rotation, H is the frame stroke and p is the tooth pitch.

To use a constant value, the root mean square tooth passage frequency may be used, which is roughly 70% of the maximum value calculated from Eq. (6.18). It is supposed that the cyclic washboarding profile is generated more by forced vibration

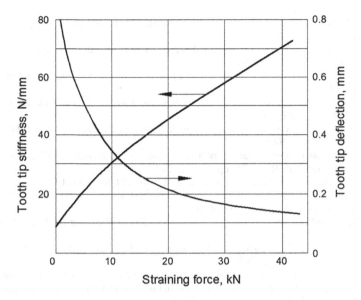

Fig. 6.30 Tooth tip stiffness and deflection as a function of the strain force. $L = 1$ m, b = 120 mm, $s = 2.2$. mm, $h = 15$ mm

of the blade due to lateral forces. However, if the ratio of the mean tooth passage frequency and natural frequency of the blade is close to one, then the height of waviness increases.

Frame saw blade can also be roll-tensioned to improve the stiffness of the tooth region using three symmetrically spaced roller passes (Bajkowski 1967).

Literature

Bajkowski, J.: Spannugnsverteilung in durch Walzen vorgespannten Gattersägeblättern. Holztech-nologie 258–262 (1967)

Csanády, E., Kovács, Z., Magoss, E., Ratnasingam, J.: Optimum Design and Manufacture of Wood Products, p. 421. Springer, Heidelberg (2019)

Lister, P., et al.: Experimental Sawing Performance Results. In: Proceedings of the 13th IWMS Vancouver, pp. 129–147 (1997)

Mote, C.: Free vibration of initially stressed circular disk. Trans. ASME 258–264 (1965)

Münz, U., Thiessen, B.: Straightening and tensioning of high hardness circular saw blades by Laser Beam. In: Proceedings of 16th IWMS, Matsue, pp. 234–247 (2003)

Münz, U.: Vibration behaviour and residual manufacturing stresses of circular saw blades. In: Proceedings of 17th IWMS, Rosenheim, pp. 407–417 (2005)

Orlowski, K., Wasielewski, R.: Washboarding during cutting on frame sawing machines. In: Proceedings of 15th IWMS Los Angeles, pp. 219–228 (2001)

Pahlitsch, G., Friebe, E.: Über das Vorspannen von Kreissägeblättern. Holz als Roh- und Werkstoff **31**, 457–463 (1973), 5–12 (1974)

Pahlitsch, G., Putkammer, K.: Beurteilung für die Auslenkung von Bandsägeblättern. Holz als Roh- und Werkstoff **32**, 295–302 (1974)

Pahlitsch, G., Putkammer, K.: Beurteilung für die Auslenkung von Bandsägeblättern. Holz als Roh- und Werkstoff **34**, 413–426 (1976)

Renshaw, A.: Centripetal tensioning for high speed circular saws. In: Proceedings of 14th IWMS Paris, pp. 129–135 (1999)

Stakhiev, Y.M.: Today and tomorrow circular sawblades: Russian version. Holz als Roh- und Werkstoff **58**, 229–240 (2000)

Stakhiev, Y.: Research on circular saw tensioning in Russia. In: Proceedings of 15th IWMS Los Angeles, pp. 293–302 (2001)

Sugihara, H.: Theory of running stability of band saw blades. In: Proceedings of 5th Wood Mach. Sem. UC For. Prod. Lab. Rich., CA. pp. 99–110 (1977)

Swift, H.W.: Cambers for belt pulley. Proc. Inst. Mech. Eng. **122**, 627–659 (1932)

Taylor, J., Hutton, S.: Simplified bandsaw tensioning procedure for improve blade stiffness and sawing accuracy. For. Prod. J. 38–44 (1995)

Weaver, W., Timoshenko, S., Young, D.: Vibration Problems in Engineering. Wiley, New York (1990)

Wong, D., Schajer, G.: Effect of wheel profile in bandsaw tracking stability. In: Proceedings of 13th IWMS Vancouver, pp. 41–52 (1997)

Маталин, А.А.: Причины возникновения остаточных напряжений. В кн. »Качество обработанных поверхностей». (Origin of residual stresses. In: Quality of machined surfaces.) Лонитомаш. Кн. 34 М. – Л. Машгиз. 1954

Санев, В.: Обработка древесины круглими пилами.(Woodworking with circular saws.) Изд. Лесная Пром., 1980

Стахиев, Ю., Определение оптимального расположения ширин зоны вальцевания дисков пил (Optimum rolling radius and width for circular saws). Изд. Лесная Пром. 1966, с. 23–25

Chapter 7
Tool Wear

7.1 Introduction

Woodworking tools generally operate at high speeds (up to 50–80 m/s), their working surfaces (rake face, edge and clearance face) are subjected to large pressures due to cutting forces. The high speed motion generates high friction power turning into heat. Therefore, woodworking tools operate under heavy mechanical and thermal loads.

The cutting process is always associated with tool wear. Tool wear is actually a loss of material from the cutting edge due to mechanical, thermal and chemical effects. The wear process leads to bluntness of the cutting edge, which in turn causes an increase in power consumption, feed force and surface deterioration of the workpiece. Surface roughness and force components are especially sensitive to changes in edge profile geometry due to wear.

The appropriate tool in the woodworking industry achieve a high wood machining capacity, high dimension accuracy, good surface roughness and long service life. The costs allocated to buying and maintaining tools is significant in relation to the total production costs. Therefore, it is important to make the optimum selections of tools for different woodworking operations, and to use them rationally to ensure a long service life.

Tool wear is a very complex process due to the large number of constituents in wood and the presence of water and all of its details are not yet fully clear. Therefore, most of our knowledge concerning wear regularities and relationships are supplied by measurement results.

E. Csanády and E. Magoss, *Mechanics of Wood Machining*, https://doi.org/10.1007/978-3-030-51481-5_7

7.2   Edge Profile and Change in Cutting Force

The initial edge profile is generally produced by grinding. Depending on the properties of the material, (first of all its hardness,) this profile is never smooth and has an irregular geometry. The initial profile of the tool (concerning its stress state) edge is not smooth enough, and the form is not optimal. After a tool is put into use, its edge profile undergoes a rapid change and takes on the shape of a smooth cylinder. The wear processes cause the radius of the tool edge to continuously increase and several stages of the tool wear can be distinguished. The initial edge radius, depending on the hardness of the tool material, is generally to 10–15 μm. As the initial stage of tool wear developed, the optimum edge shape extends to an approximate radius of 20 μm (Fig. 7.1).

The next and more important stage is the "working sharp" stage, which ensures the optimum cutting conditions with low energy consumption and good surface quality. Over an edge radius of 40 μm, the tool starts to become blunt, requiring increasing cutting forces and decreasing the surface quality. The tool in this stage of wear is suitable for rough woodworking operations, but it does not ensure a good surface quality. Finally, in the blunt wear stage the tool requires high cutting energy and produces low surface quality. In addition, the blunt edge compresses the upper layers of the cut surface causing severe damage and surface instability. Furthermore, the high radial force component, as an exciting force, causes vibration with high amplitudes worsening the surface quality considerably.

Quantitative measurement of tool wear is of great importance to the woodworking industry. Wear can be defined as a loss of material from the cutting edge due to mechanical, electro-chemical and high temperature effects associated with the cutting process. The direct measurement of material losses would be complicated. Therefore, different parameters of the edge profile are used. The shape of the worn edge may be

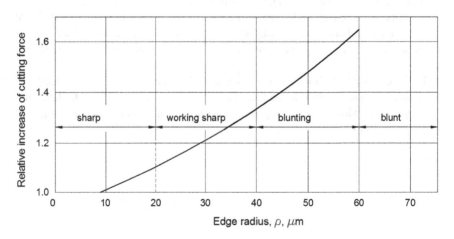

Fig. 7.1 Stages of tool wear and the relative increase of the required cutting force

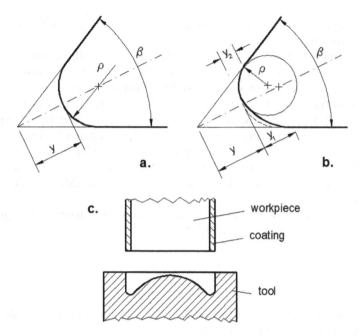

Fig. 7.2 Different edge wear profiles: **a**—symmetric, **b**—asymmetric, **c**—wear due to workpiece coating

symmetric or asymmetric. In the simpler symmetric case, there is a uniquely defined relationship between the edge radius ρ and the initial position of an ideally sharp cutting edge y (Fig. 7.2).

$$\rho = y\frac{\sin\left(\frac{\beta}{2}\right)}{1 - \sin\left(\frac{\beta}{2}\right)} \tag{7.1}$$

where β is the sharpening angle of the tool edge.

Due to the higher pressures acting on the bottom (clearance) face of the edge, the wear profile generally assumes an asymmetric shape (Fig. 7.2b) allowing other measurement parameters to be used. A cylindrical edge surface will also be developed here, but its centre point lies slightly over the bisector.

Particleboard and coated wood materials have an uneven density and hardness distribution; therefore, their wear action can be quite different along their cross section (Fig. 7.2c). Some glues may have strong abrasive properties.

The most commonly measured wear parameter is edge retraction y, which can easily be measured. This single wear parameter does not provide full information about the shape and geometry of a worn edge. It was also observed that the nose radius

(Fig. 7.2b) may be stabilized at a particular value and does not always increase further with increasing wear.

It is also important to note that the characteristic wear profile shown in Fig. 7.2b remains similar using different tool materials, such as tool steel, high speed steel and carbide tools. That also applies to dry and green wood or particleboard.

The quantifying of tool wear when the wear varies along the cutting edge is more complicated. One possible method is to calculate the area under the wear profile along the edge. A simpler way is to consider the wear only below the outer surfaces of the board where the density is the highest and the maximum wear of the tool occurs. It is found that nose (or rake) recession has a good correlation with the area under the nose recession curve drawn along the cutting edge (Sheikh-Rahmad and Mckenzie 1997).

Tool wear can be related to the feed distance (L_f) and the true cutting length (L_c). The latter can be calculated as (see Fig. 7.3).

$$L_c = L_f \frac{R \cdot \varphi}{e_z \cdot z} \tag{7.2}$$

where R is the tool radius, e_z is the tooth bite, z is the number of teeth and the angle ot cutting φ must be substituted in radian.

The true cutting length when peeling wood will simply be the veneer length produced.

Fig. 7.3 Calculation of true cutting length

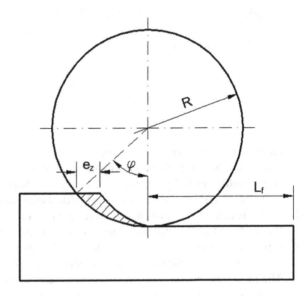

7.3   The Major Tool Wear Mechanisms

Three major wear mechanisms can be distinguished in tool wear: abrasion, electro-chemical and high temperature corrosion. Generally, abrasion is always present as a mechanical wear process acting together with one of the other two, depending on the density, moisture content, and acidity of the wood and cutting parameters, especially edge surface temperature; (see in Sect. 2.4).

Abrasion occurs from the motion of hard particles between the sliding surfaces of a tool and wood. Wood species have always some mineral contaminations, which can be estimated by using a combustion method. The remaining hard mineral contaminations can be separated into fractions and measured. Their particle size is generally smaller than 50 μm. Silica is especially responsible for excessive blunting of tools. The silica content may be very different ranging from 0.1 to 10 g/kg wood.

The density of wood is also important in the process of mechanical wear. The denser wood has higher mechanical strength and exerts higher pressure on the edge face of a tool.

The amount of abrasive wear is proportional to the length of sliding contact between the tool and workpiece:

$$Q = A \cdot l$$

where A is a constant.

The rate of wear due to abrasion is given by differentiating the above equation

$$\frac{dQ}{dt} = A\frac{dl}{dt} = A \cdot v$$

This equation shows that the rate of wear is linearly related to the cutting speed.

Electro-chemical wear is attributed to the extractives in wood, such as gums, resins, sugars, oils, starches, alkaloids and tannin in the presence of water. Most extractives are reactive compounds forming chemical reactions between the wood extractives and the metallic constituents of a tool. The acidity of wood (pH-number) has great importance in this wear process. The cutting process is always associated with electric discharges contributing to the wear process. Early experiments have shown that negative electrical potentials imposed between the tool and workpiece can inhibit tool wear (Kivimaa 1952; Alekseyev 1957), while positive electric potentials increased the wear rate.

The amount of electro-chemical wear depends on the length of time that the tool-electrolyte interaction occurs and it can be described by the following simple equation:

$$Q = B \cdot t$$

The wear rate is constant for a given combination of tool and workpiece:

$$\frac{dQ}{dt} = B$$

The constant B depends on wood species, moisture content and especially on its acidity (Tsai and Klamecki 1980; Packman1960). For example, oak and incense cedar (Calocedrus) have low pH-values.

The metal components in a tool also have a definite effect on the wear rate. High alloyed *HSS* steel is more susceptible to chemical attack from the extractives than WC and Co in tungsten carbide.

High temperature corrosion is also a common wear mechanism due to high temperatures in the cutting zone. High temperatures around 800–900 °C may occur depending on the cutting speed and other operational parameters. This high temperature can induce oxidation, which attacks the tool material. Different wood species have different behaviour in high temperature oxidation. The intensity of high temperature corrosion also depends on tool material and wood species. Using cemented tungsten carbide tools, it was found that the main corrosion mechanism is the oxidation of the cobalt binder which deteriorates the structure of the material in the surface layer. Hot corrosion is generally associated with the formation of salt deposits on the surface. A close examination of these deposits has revealed that wood constituents, such as *Ca* and *K* are present on the knife edge. Both impurities *Ca* and *K* are known to accelerate hot corrosion processes. Furthermore, the deposits always contain a considerable amount of carbon, which also accelerate hot corrosion.

Making electron micrographs of *HSS* tool edges showed that apparent smearing and plastic deformation are present indicating high temperatures and pressures which are contributing factors to mechanical wear (Stewart 1989).

Each of the wear processes described above may affect tool wear to a varying degree depending upon the specific machining conditions. They may interact among themselves or be part of other processes. For example, additional abrasion may result from oxides, carbides, nitrides and other compounds formed from high temperature corrosion processes.

Considering high temperature wear, one should also take into account the effect of adhesives, finishes and other materials used to improve properties of wood products. Chloride and sulphate salts found in adhesives accelerate hot corrosion. It is not surprising that the machining of particleboard containing adhesive materials is always associated with high wear.

7.4 Factors Affecting Wear

To reduce tool wear, the selection of appropriate tool materials or their treatment, such as coatings, is very important. Coatings are more resistant to high temperature corrosion and provide a longer tool life. Polycrystalline diamond is highly resistant to abrasion and to chemical attack. A high coefficient of heat conduction in a tool material is of great importance in lowering the surface temperature in the cutting zone. A larger sharpening angle also lowers the maximum edge temperature; (see in Sect. 2.4).

Edge reduction due to wear can be expressed by the following empirical equation:

$$y = y_0 + A \cdot v^n \cdot L_c^m \tag{7.3}$$

where y_0 is the initial edge reduction after grinding, v is the cutting speed. The constants A, n and m can be determined experimentally.

The total feed distance during the life time T of a tool is:

$$L_{fT} = e_z \cdot n \cdot z \cdot T$$

where e_z is the tooth bite, n is the rotation speed and z is the number of teeth, while the true cutting distance during its life time is:

$$L_{cT} = n \cdot R \cdot \varphi \cdot T = \frac{60}{2\pi} v \cdot \varphi \cdot T \tag{7.4}$$

where R is the radius of the tool.

The wear and lifetime of the tool is determined by cutting distance and not by the feed distance. The use of feed distance is, however, more convenient in many cases and, therefore, the L_c/L_f ratio is important:

$$\frac{L_c}{L_f} = 1.42 \frac{\sqrt{H \cdot R}}{e_z}$$

or

$$\frac{L_c}{L_f} = 1.42 \cdot \frac{n \cdot z}{e} \sqrt{H \cdot R}$$

where
    H is the depth of cut,
    The rotation angle of cutting φ was substituted as

$$\varphi = 1.42 \cdot \sqrt{\frac{H}{R}}$$

For example, using $R = 60$ mm, $H = 2$ mm and $e_z = 1$ mm this ratio is $L_c/L_f = 15.61$.

Using Eq. (7.3) and expressing L_C, we obtain

$$L_c = \left(\frac{y - y_0}{A}\right)^{\frac{1}{m}} \cdot \frac{1}{v^{\frac{n}{m}}},\tag{7.3a}$$

Taking a wear limit, of 100 μm, and combining Eqs. (7.4) and (7.3a), the following general expression is valid (Taylor tool life equation):

$$T \cdot v^{\frac{n}{m}+1} = \text{constant}\tag{7.5}$$

or

$$v \cdot T^{\frac{1}{\frac{n}{m}+1}} = \text{constant}\tag{7.5a}$$

Figure 7.4 shows experimental results for particleboard using a tungsten carbide tool ($\beta = 55°$, 1.2 mm tooth bite, 2–3 mm cutting depth, 60 mm tool radius) (Saljé and Dubenkropp 1983; Saljé and Drückhammer 1984).

The initial edge recession is $y_0 = 20\,\mu m$ ($\rho = 17\,\mu m$) and the calculated exponents in Eq. (7.3) are $n = 0.74$ and $m = 0.65$. Using these exponent values, the Taylor equation assumes the form

$$T \cdot v^{2.14} = const$$

or

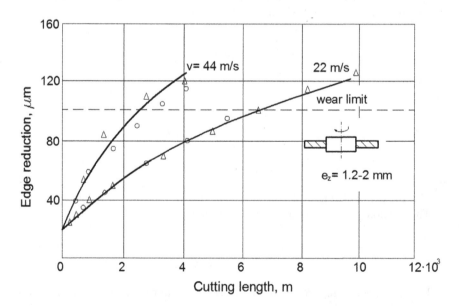

Fig. 7.4 Edge reduction as a function of cutting length for different cutting velocities

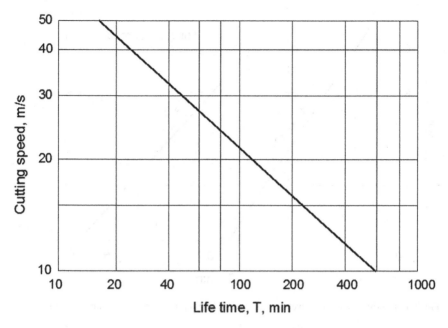

Fig. 7.5 Taylor life time curve (Based on data in Fig. 7.4)

$$v \cdot T^{0,47} = const$$

Using logarithmic co-ordinates, the Taylor equation is represented by a straight line (Fig. 7.5), which can easily be used to determine the service life after re-sharpening a tool.

The maximum cutting length during the service life of a tool is (see Eq. (7.3a)):

$$L_{cT} = \frac{const}{v^{\frac{n}{m}}}$$

The exponent $n/m = 1.14$ in Fig. 7.4 shows that with increasing cutting speed, the maximum cutting length decreases. The exponent in the Taylor tool life equation has different values depending on the dominant wear process (abrasive, electro-chemical or hot corrosion) in a particular case. In pure abrasion, the exponent of T is equal to one; for electro-chemical wear, this exponent is greater than unity and for high temperature wear, it is less than unity. The maximum cutting length during the service life of a tool may be independent or dependent on the cutting speed.

Figure 7.6 shows Taylor tool life diagram for Ponderosa pine in the low cutting velocity range (Tsai and Klamecky 1980).

The exponent of T is greater than unity depending on the moisture content. That means that the dominant wear mechanism is the electro-chemical wear. The low cutting speed ensures moderate heat generation on the tool edge avoiding high

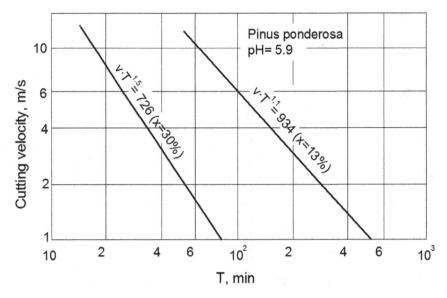

Fig. 7.6 Tool life diagram for wet and air-dry Ponderosa pine in the low cutting speed range

temperature corrosion. Cutting air-dry Ponderosa pine, however, the value of exponent is 1.1 quite near to the unity. Therefore, abrasion seems to be the dominant wear mechanism in this case. As a consequence, the maximum cutting length during the service life is more or less independent of cutting speed.

The tool life equation can be written in an other form. Combining Eqs. (7.4) and (7.3a) yields

$$T \cdot v^{\frac{n}{m}+1} = \left(\frac{y - y_0}{A} \right)^{1/m} \cdot \frac{0.1047}{\varphi} \tag{7.6}$$

which is one form of the Taylor tool life equation and for a given case the right side of Eq. (7.6) is constant. Using the value of constants in the above equation as $A = 0.0245$, $m = 0.78$ and $n = 0.65$, the Taylor tool life equation can be written as

$$T \cdot v^{2.2} = 91957 \text{ or } v \cdot T^{0.4545} = 180.4$$

which is plotted in Fig. 7.7.

The constant on the right side depends on the rotation angle of cutting φ which is the function of the relative cutting depth as given formerly.

Knowing the relation between edge reduction y and edge radius ρ (appr. $\rho = 0.85y$), in Eqs. (7.3a) and (7.6) ρ and ρ_0 can also be used.

It is interesting to note that the true cutting length does not depend on the depth of cut H but the feed distance does. In practice the use of feed distance is more convenient and it can be expressed using Eq. (7.6) for T

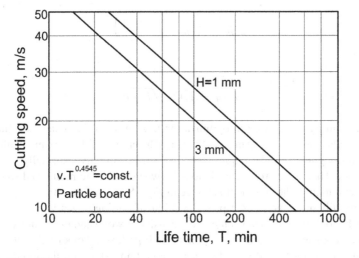

Fig. 7.7 Taylor lifetime curve

$$L_{fT} = e_z \cdot n \cdot z \left(\frac{y - y_0}{A} \right)^{1/m} \cdot \frac{0.1047}{\varphi \cdot v^{\frac{n}{m}+1}} \qquad (7.7)$$

The cutting speed v is directly proportional to the rotation speed n and, therefore, the feed distance is inversely proportional to $v^{n/m}$ which is depicted in Fig. 7.8 using $e_z = 1$ mm, $z = 4$ and $R = 60$ mm.

The depth of cut considerably influences the feed distance. Changing the rotation speed into cutting speed, Eq. (7.7) takes the form

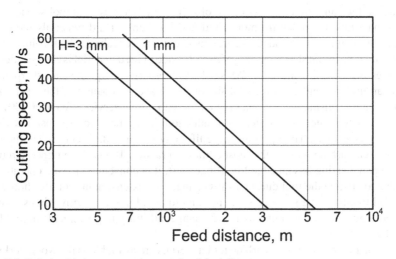

Fig. 7.8 Relationship between cutting speed and feed distance

$$L_{fT} = \frac{e_z \cdot z}{R \cdot \varphi \cdot v^{n/m}} \cdot \left(\frac{y - y_0}{A}\right)^{1/m} \text{ m} \qquad (7.7a)$$

with

$$R \cdot \varphi = 1.42 \cdot \sqrt{H \cdot R}$$

which can easily be used in optimization too. In the above examples, the wear limit was taken as $y_0 = 100 \ \mu m$ ($\rho = 85 \ \mu m$). The choice of the maximum allowable wear limit is guided by the requirements for surface quality, especially roughness parameters. There are strong correlations between tool edge radius and roughness parameters, mainly the core depth R_k (see in Figs. 8.36, 8.37, 8.38).

Excess vibrations and circle path deviation of cutting edges will considerably shorten the tool life. The resin content, the density of wood and the quantity of adhesives in composite boards alter the exponents in Eq. (7.6) and modify the relationship between cutting speed and feed distance to a certain extent. Furthermore, the tool material (carbide grain size) may also affect the wear characteristics (Sugihara et al. 1979; Salje and Dubenkropp 1983).

Tools require a careful maintenance to ensure an accurate edge geometry and running circle. Another requirement is to avoid excessive heat loads when grinding which may cause permanent tensile stresses in the surface and undesirable changes in the structure of the metal. This means an abnormal wear process is taking place rapidly blunting the edge.

The most common tool materials used are the low alloy steel, *HSS* steel, stellite tipped tool, cemented tungsten carbide, boron nitride and polycrystalline diamond. Various edge treatment and coating procedures were examined for further increase of the service life of tools (Gottlob and Ruffino 1997). The above list of tool materials means, at the same time, the sequence of increasing hardness and tool service life. Hard materials are generally more rigid and breakable. Hard materials require a bigger sharpening angle and, as a consequence, a smaller rake angle. It is important to note that the initial sharpness (y_0) of a tooth made of hard material is always inferior to edges made of low alloy or *HSS* steel. Peeling knives require small sharpening angle around 20° and a good initial sharpness. A special coating of the steel knife will maintain satisfactory sharpness of a tool with a long service life (Beer 2001).

In order to achieve cost-effective production, it is crucial to use cost and edge life optimised tools. The optimum edge life may be quite different. Today the use of tungsten carbide tools is the most common practice. If a carbide tipped circular saw in panel sizing has an edge life of one shift, it is convenient for the operator to be able to change the tool during a shift change. An extension in tool life should be doubled to ensure the same convenience and to avoid non-productive times. When the working time is not bound to a shift change, arbitrary improvements in tool life may be used.

A wide range of tungsten carbide tool materials are available to the woodworking industry for many different uses. The different carbide grades are composed of various percentages of tungsten carbide and a metallic binder (cobalt). Furthermore, these tool materials utilize a range of carbide grain sizes. The most common grain size is about 2 μm but there are grades in the sub-micron range (Feld et al. 2005; Garcia 2005). Varying the carbide-binder ratio and the grain size, tool materials with different mechanical properties are made to ensure optimum performance when machining different types of wood and composites (hard and softwoods, MDF, particleboard etc.).

Composite materials require hard tool materials, with sharpening angles of around 60° or more. This requires carbide grades with less cobalt binder (2.5–3%) and small carbide grain size (0.5–0.7 μm). For soft and hardwoods, grades with high binder content (around 10%) with grain size of 0.7–1.5 μm with good fracture resistance are the best choice which enable operators to use a smaller sharpening angle.

To cut veneer, a knife with a sharpening angle of around 20° is used. The small sharpening angle requires high-speed steel (HSS) which is generally hard coated with chromenitride (CrN) or nitrided in a gas mixture of nitrogen, oxygen and methane, (a duplex treated knife) (Chala et al. 2003). An optimum treatment increases the service life of the tool by a factor of near 2. Due to the treatment, the friction coefficient of the tool may be lowered considerably. Furthermore, the lifetime can also be extended by using a back micro-bevel angle. A smaller sharpening angle removes less heat from the edge (see Fig. 2.10). As a consequence, one should reckon with an increasing wear rate resulting in a higher exponent m in Eq. (7.6). A further condition for an adequate lifetime is a uniform running circle of the edges (Tröger and Läuter 1983). Deviations in the running circle means different tooth bites for certain edges resulting in varying surface waviness and excessive vibrations due to self-exciting which may further enhance unequal tooth bites. A possible solution is to use jointing which minimises the running deviation of edges. Jointing may have another beneficial effect. It makes the edge sharper and, decreases the compaction of neighbouring layers due to the cut of bottom convex part of the edge. As a consequence, it reduces the surface roughness, especially the core roughness component R_k.

An interesting suggestion is to create a "self-sharpening" edge, whose principle has been used for a long time in agricultural engineering as the self-sharpening ploughshare (Itaya and Tcuchiya 2003). A thin layer of about 10 microns made of chromenitride is laid down on the clearance face of the knife. Since coated layer is harder than the base metal, it wears more slowly and the edge will be kept sharp due to the thin hard layer. The knife of a Japanese planer is also covered with a thin layer of special steel, which can be sharpened to an extremely small edge radius.

Diamond and diamond coated tools are also in use in the wood industry for heavy woodworking operations. They are expensive and sensitive to breakage and therefore their economic use requires detailed considerations.

Finally, the main factors influencing tool wear can be summarized as follows:

7.4.1   Tool Properties

- tool material
- sharpening angle
- rake angle
- edge treating or coating
- accuracy of edge running circle
- accuracy of side running.

7.4.2   Workpiece Properties

- strength properties
- moisture content and pH-value
- density
- cutting direction
- resin and ash content
- adhesives and surface coating.

7.4.3   Operational Parameters

- cutting speed
- tooth bite
- cutting depth
- vibration of the tool.

The number of influencing factors is very high, the interaction of different wear mechanisms is very complicated and, therefore, the tool life for a particular case can only be determined experimentally. In practice it can often be observed that more frequent re-sharpening of the tools is carried out to secure good surface quality under any circumstances. It is clear that the combine requirement of quality and economy can be achieved by compromise in the wise choice of influencing factors.

Literature

Beer, P.: Synergism of the abrasive and electro-chemical wear of steel tools in wood peeling. In: Proceedings of 15th IWMS Los Angeles, pp. 103–108 (2001)
Chala, A. et al.: Duplex treatment based on the combination of ion nitriding and PVD process. In: Proceedings of 16th IWMS Matsue, pp. 471–479 (2003)
Feld, H. et al.: Carbide selection for woodworking tooling. In: Proceedings of 17th IWMS, Rosenheim, pp. 520–527 (2005)

Garcia, I.: New developments in ultrafine hardmetals. In: Proceedings 17th IWMS, Rosenheim, pp. 534–542 (2005)

Gottlob, W., Ruffino, R.: AISI D-6 Steel and stellite-1 tool life determination. In: Proceedings of 13th IWMS Vancouver, pp. 761–769 (1997)

Itaya, S., Tcuchya, A.: Development of CrN coated tool to create smooth surface of solid wood. In: Proceedings of 16th IWMS, Matsue, pp. 74–81 (2003)

Kivimaa, E.: Was ist die Abstumpfung der Holzbearbeitungswerkzeuge? Holz als Roh- und Werkstoff, S. 425–428 (1952)

Packman, D.: The acidity of wood. Holzforschung, S. 179–183 (1960)

Saljé, E., Dubenkropp, D.: Das Kantenfräsen von Holzwerkstoffplatten. Holz- und Kunstoffverarbeitung. No. 4., SW. 490–494 (1983)

Saljé, E., Drückhammer, J.: Qualitätskontrolle bei der Kantenbearbeitung. Holz als Roh- und Werkstoff, S. 187–192 (1984)

Sheikh-Ahmad, J., McKenzie, W.: Measurement of Tool Wear in the Machining of Particleboard. In: Proceedings of 13th IWMS Vancouver, pp. 659–670 (1997)

Stewart, H.: Feasible high-temperature phenomena in tool wear. Forest Prod. J., 25–28 (1989)

Sugihara, H. et al.: Wear of tungsten carbide tipped circular saws in cutting particle board. Wood Sci. Techn., 283–299 (1979)

Tröger, I. Läuter, G.: Zum Einfluss von Rundlaufabweichungen auf Bearbeitungsqualität und Standvorschubweg beim Fräsen. Holztechnologie, S. 24–29 (1983)

Tsai, G., Klamecki, B.: Separation of abrasive and electrochemical tool wear mechanisms in wood cutting. Wood Sci. 4, 236–241 (1980)

Алексеев, А.: Влияние электрических явлений на износ инструментов (Effects of electric phenomena on tool wear) Дерев. Пром., 8, с. 15–16 (1957)

Chapter 8
Surface Roughness

8.1 Introduction

Machined surfaces—regardless if they are made of metal, plastic or wood—are never perfectly smooth; we can observe protruding parts, valleys and peaks on them. These forms of surface irregularity are called *roughness*. Surface roughness can be caused by different factors: discontinuities in the material, various forms of brittle fracture, cavities in the texture (e.g. wood), radius of the tool edge, local deformations deriving from the free cutting mechanism.

Surface roughness usually has a primary influence on the visual appearance of materials but it might have other effects, too. Surface roughness may be extraordinarily detrimental for wood if the surface under the tool edge suffers permanent deformity. The stability of the damaged surface diminishes largely; the durability of the processed surface will then be inferior.

Visual appearance and colour effects are primarily influenced by dispersion and reflection of light. An apparent example for this is transparent glass, which—following a moderate roughening process—loses transparency. The original colour of wood becomes a lot more visible if the surface is 'bright': that is to say smooth and free of irregularities. Speaking of wood, a good example in this context is ebony: the black colour gives entirely different effects depending on the surface roughness. The polished surface presents a bright black colour. A surface treated with colourless lacquer, oil or wax will lead to quite different colour effects again.

The minimum surface roughness that can be achieved depends on a number of factors. Generally we can say that processing materials with bigger volume weight can result in smoother surfaces. This explains the excellent polishing ability of ebony.

Among conifer species larch has usually the biggest volume weight and accordingly it is easy to machine from the aspect of surface quality. The Tasmanian Huon pine (Lagarostrobos franklinii), which is lesser known in Europe, grows extremely slowly and has an annual ring width of 0.1–0.2 mm. Its resin content is high and it can be polished excellently.

E. Csanády and E. Magoss, *Mechanics of Wood Machining*,
https://doi.org/10.1007/978-3-030-51481-5_8

The surface roughness of wood results from multiple factors, therefore defining general rules has taken quite a long time (Schadoffsky 1996; Devantier 1997; West-kämper 1996). New ideas in the latest decade and modern measurement techniques supported the identification of essential rules. We present here a summary of 25 years research work conducted at the University of West Hungary, Sopron, to establish the basic rules concerning the surface roughness of solid woods (Magoss and Sitkei 2000, 2001; Csiha 2003). Former difficulties were lying in that the wood species, as a variable, can not be expressed in terms of numerical values. In order to solve this problem it was necessary to elaborate and introduce a new *wood structure charac-terisation* method. The newly defined anatomical structure number made it possible to treat all wood species in a common system facilitating the recognition of general regularities.

This chapter discusses all aspects and the main relationships of surface roughness in a unified approach. Roughness parameters used in the wood industry, the charac-terization of internal structure of wood, the origin of surface roughness are treated in detail. Effect of machining and internal structure on roughness and interrelation between roughness parameters are highlighted. The till now less used skewness and kurtosis are introduced and discussed.

8.2 Measurement Methods

First the two-dimensional, mechanical stylus measuring method (perthometer) was developed which is used also today. This reliable method uses a fine needle (stylus) with tip radius of 2–5 μm which is capable to scan quite small irregularities on the surface. At the same time, the stylus, due to its geometrical dimensions (edge radius, cone angle), performs some so-called mechanical filtering. Furthermore, the stylus is unable accurately to scan all cavities to the full depth. The measurement principle is shown in Fig. 8.1 (Magoss 2008).

The pressing force of the stylus has decreased to a value around 0.8–7 mN by today, thus it has no detrimental effects on the surface. During testing on a Scotch pine probe no surface damage was experienced at a 40-times repetition (Westkämper 1996). However, there is a special case when even this minimal pressing force could become problematic: if the stylus contacts a free fibre—that is still connected to the surface at its other end—in a perpendicular direction. In this case the stylus simply pushes the fibre aside. The same phenomenon brings an advantageous effect when the stylus meets a particle of dust on the surface. The working principle of perthometers is shown in Fig. 8.1.

The diamond stylus tip is installed with a suspension of minimal friction resis-tance. While the stylus is drawn at a constant speed, an electromechanical converter (differential transformer) converts the vertical shift of the stylus into an electric voltage (see Fig. 8.1).

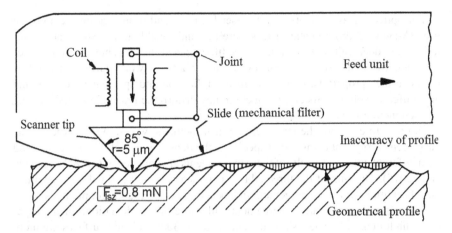

Fig. 8.1 Installation of the stylus

The signal is first amplified and then evaluated (Fig. 8.2). For the purpose of evaluation, roughness is separated from waviness by means of frequency filtering. A detailed discussion of frequency filtering will follow later on. Some manufacturers enable the installation not only of a mechanical micro stylus system on the feed but also that of an optical micro stylus system; for instance the instrument 'Focodyn' by Mahr which uses the laser focusing principle.

In the last decades an increasing need has arisen for more detailed characterization of surface topography in three dimensions (3D). Some of the 3D parameters are extended from the previously defined 2D counterparts, while others are newly defined. The proposed parameter set was mainly determined to meet the demands of the metal industry (Thomas 1981; Dong et al. 1994). Optical devices are mainly used to perform 3D measurements.

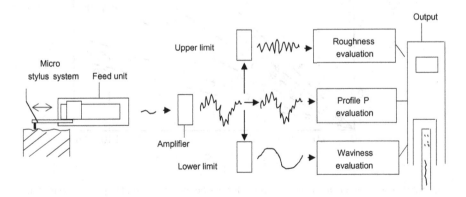

Fig. 8.2 The working principle of perthometers

For optical measurements mainly laser focusing and triangulation methods are used (Magoss 2008). Optical methods are fast and capable to measure fine irregularities and deep valleys too. At the same time, optical methods have serious disadvantages, especially when measuring wood surfaces. A wood surface, depending on its surface properties, may have diffuse reflection and specular reflection, and also different colours. Specular reflection may disturb the receiving signal causing measurement errors.

In order to compare the stylus and optical methods, wide-ranging experiments have been carried out using wood species in the density range of 0.3–1.2 g/cm^3 and with different colours. Results of these experiments are depicted in Fig. 8.3 (Csanády et al. 2015). The sum of the three Abbott parameters is well defined and they must have been more or less in agreement independently of measurement methods.

While some species are in agreement with more or less deviation, others, especially high-density species, show deviations up to 300%. Planed surfaces are more prone to deviations due to their higher reflectivity. Until this problem is solved, it is wise to check the validity of optical measurement results.

A special surface quality problem is the edge breaking and its proper numerical characterization. Narrow face machining of panels and other components is often required and an important requirement is to produce unbroken, sound edges. The

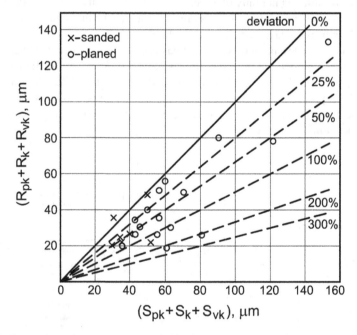

Fig. 8.3 Comparison of measurement results for 2D Perthometer and 3D optical system. Each point represents a given wood species

broken part is characterized by the three orthogonal dimensions l_x, l_y, and l_z, and the number of breaks related to the unit length, Fig. 8.4. The edge is traced with a wedge set under a certain angle α. The wedge measures the depth and length of the breaks, the width can approximately be calculated using the depth and angle α.

The essence of evaluation of the measurement results is demonstrated in Fig. 8.5 (Saljé and Drückhammer 1985). Several depth intervals are defined and the broken cross-section falling into the given interval will be weighted with an increasing factor C_i as a function of depth. The characteristic number Q_A is calculated as follows

$$Q_A = \frac{1}{L} \cdot \sum \Delta A_{ij} \cdot C_i \ \text{mm}^2/\text{m} \tag{8.1}$$

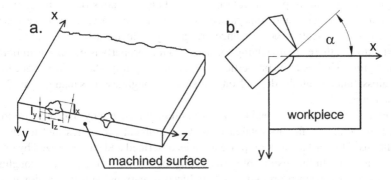

Fig. 8.4 Edge breaking characterization (**a**) and its measurement principle (**b**)

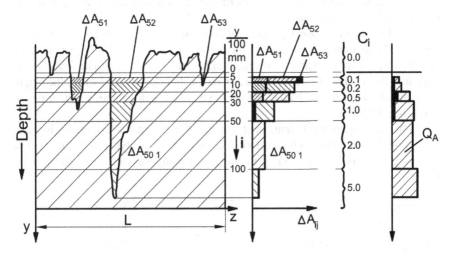

Fig. 8.5 Evaluation of the measurement results (Saljé 1984)

where

L is the trace length,

i is the number of depth intervals ($i = 6$),

j is the number of breaks in the trace length.

There is a threshold (0.05 mm) below of which the cross-sections are not counted ($C_i = 0$), due to their insignificant effect on edge quality. On the contrary, deeper breaks considerably worsen the edge quality.

8.3   Parameters of Surface Roughness

Roughness plays an important role in several fields of industrial technology with different requirements and importance. In the metal industry, the surface bearing capacity and lubrication performance are greatly influenced by roughness. In the wood industry, there are no sliding surfaces and this property is much less important or negligible. The aesthetic properties of wood are very important. The surface finish and gloss fundamentally depend on the surface roughness (Csanády et al. 2015; Richter et al. 1995).

Engineered wood surfaces have a very complex topography and they are influenced by countless factors related to their anatomical properties and machining conditions. This surface topography cannot be described and characterized by one or few parameters. In the history of surface roughness measurement, many roughness parameters have been proposed and introduced in national and international standards (Csanády et al. 2015; Dong et al. 1994; Thomas 1981; Abbott and Firestone 1933). Machined surfaces always show irregularities of different height and depth; this is what we call *roughness*.

Typical surface roughness profiles for soft and hardwood species are shown in Fig. 8.6. The first curve demonstrates the surface roughness of Scotch pine of thick growth, where both high and deep irregularities are of small size. The fourth curve illustrates the machined surface of black locust with large vessels, where the height of irregularities are below 10 μm but the large vessels cause depth irregularities of size between 50 and 80 μm.

For roughness measurements first the two-dimensional (2D) measuring and evaluating system has been developed mainly using a stylus instrument. The following roughness parameters will most commonly be used to evaluate wood surfaces:

Parameter	Symbol
Average peak-to-valley height	R_z
Maximum peak-to-valley height	R_{max}
average (integrated) roughness	R_a
Root-mean-square (RMS) roughness	R_q

(continued)

(continued)

Parameter	Symbol
Skewness (the third moment of profile amplitudes)	R_{sk}
Kurtosis (the fourth moment of profile amplitudes)	R_{ku}

The definition of R_z, R_{max}, R_a and the waviness are given in Fig. 8.7. The possible variations of skewness and kurtosis are explained in Fig. 8.8.

Since 1933 the Abbott-Firestone roughness parameters are in use which characterize the material and void distribution in a rough surface These parameters are the following (Abbott and Firestone 1933):

Parameter	Symbol
The average height of protruding peaks	R_{pk}
Vertical difference in the core section	R_k
The averaging height of valleys	R_{vk}
Material volume above the upper core surface	M_{r1}
Void volume under the lower core surface	M_{r2}

Figure 8.9 shows the Abbott-Firestone curve and the meaning of individual parameters.

A part of the space within the roughness profile is filled with material, the rest of the space is filled with air. The relationship between these two factors is expressed by the material ratio curve of the profile, as in distribution curves. The material ratio curve of the profile is also known as the Abbott curve; its definition and parameters are shown in Fig. 8.9. Surface roughness measuring instruments perform this assessment automatically; the data are drawn and can be printed.

Furthermore, optical devices restructure the Abbott parameters S_{pk}, S_k, and S_{vk}, relative to each other as compared to the appropriate R_{pk}, R_k and R_{vk} values.

Due to its definition, S_z is dominated by extreme values and it is hardly comparable with its R_z counterpart.

In contrast with other opinion, the R_q/R_a ratio is an important factor and has a strong correlation with the R_{sk} skewness. Therefore, both R_a and R_q are useful roughness parameters in the wood industry (Csanády et al. 2015).

Returning to the Fig. 8.6 it is interesting to note that only the bigger vessels are visible on the roughness profile. It may be explained by the fact that the surface layer is deformed (crushed) under the cutting edge and the machined surface is not clean. A more or less clean surface can only be made with the Japanese planer.

The Japanese planer has been used for many centuries and its ability to produce smooth surface is also well known but measurements of its effects on surface quality are not known. Samples of several wood species were planed with different chip thicknesses between 15 and 80 μm. For comparison, sanded surfaces with P-240 were also prepared. A selection of roughness profiles is shown in Fig. 8.10. The Japanese

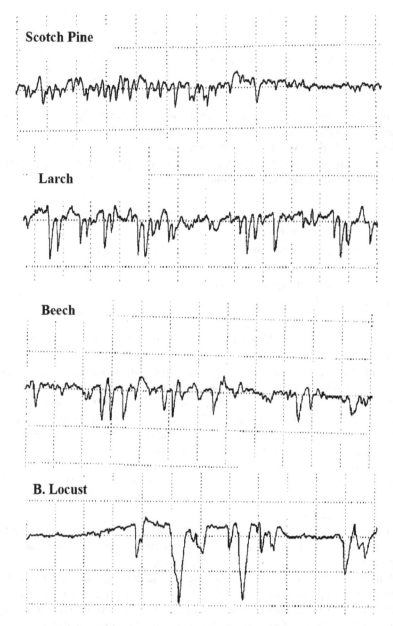

Fig. 8.6 Surface roughness profiles

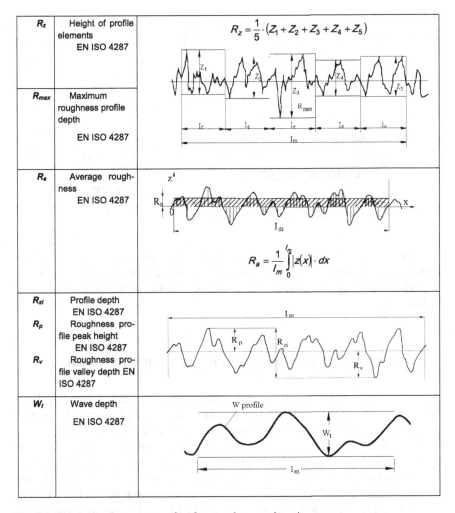

Fig. 8.7 Standardized parameters of surface roughness and waviness

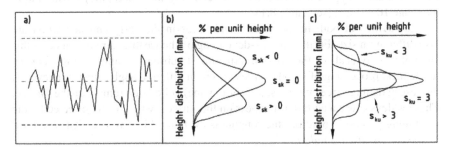

Fig. 8.8 Height distributions for skewness (**b**) and kurtosis (**c**)

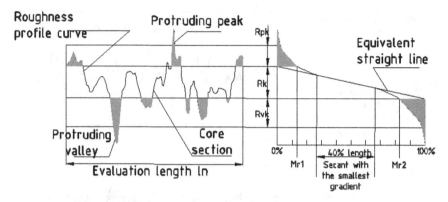

Fig. 8.9 The Abbott-Firestone curve and its related parameters

planer produces extremely clean surfaces and the repeating early and latewood are clearly seen. The same fine sanded sample shows no signs of early or latewood.

A clean surface has excellent gloss properties and its colour hue is enhanced in certain extent.

8.4   Internal Structure of Wood and Its Characterization

One of the typical characteristic features of wood is the internal structure, which has cavities in the form of vessels and cell lumens inside. The typical internal structure of soft and hardwood species is shown in Fig. 8.11 (Butterfield et al. 1997).

The early wood tracheids of Scotch pine have large cavities (20–40 μm) and thin walls, however cavities in the late wood roughly are half of the size. The structure of hardwood species is more complicated, they consist of a number of different cell types. Vessels consisting of vertical units doing the transportation, the diameter size of which can be up to 300 μm; thus they are visible to the eye. The tracheids—cavities in the long parenchyma cells—are relatively small here; they mostly fall in the range 10–20 μm (see Table 8.1)

The cavities of both in the vessels and fibres in the early and late wood are also different in size. Furthermore,—depending on the weather circumstances—this variability is typical for the internal structure of subsequent annual rings. During mechanical processing of wood cavities are cut in different angles, therefore even in the case of damage-free cutting (sharp cutting line) hollows do remain on the surface. These valleys cause a certain roughness on the surface, which is not effected by the machining process. Therefore the roughness evolving this way is called *structural* or *structure-caused roughness*.

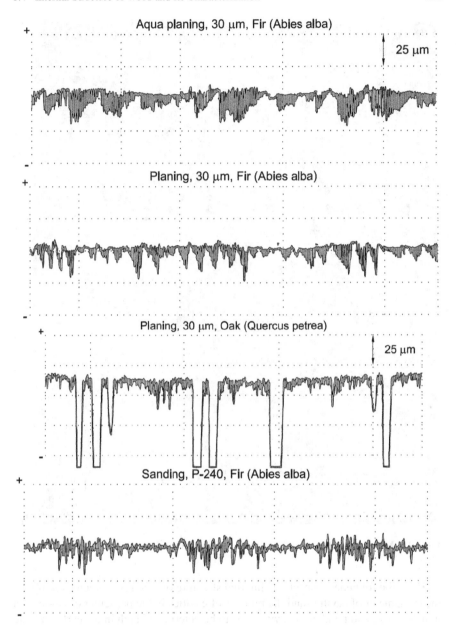

Fig. 8.10 Surface roughness profiles produced by a Japanese planer. Chip thickness is 30 μm

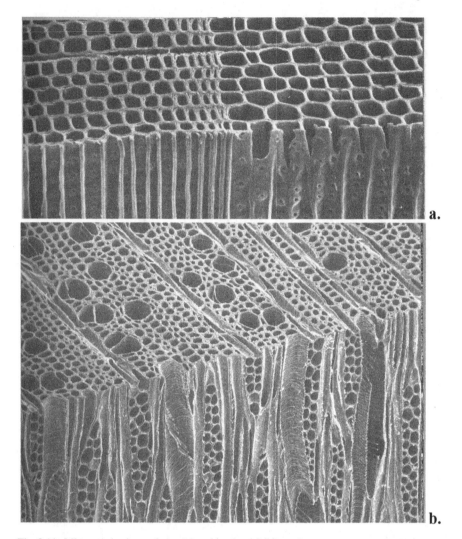

Fig. 8.11 Microscopic photo of pine (**a**) and hardwood (**b**) species

From the roughness aspect, the internal structure of wood is characterized by the mean diameter of cavities and the number of cavities in the particular cross-section unit. The size and number of cavities has to be determined both in the early and late wood, therefore, the early and late wood ratio must also be established.

In order to gain the above data, small-sized sections are taken from different wood species, where the required data are established using a measuring microscope or by means of digital image processing. It is practical to check the obtained data also by calculation methods. The bulk density of the sample is easy to determine; the following approximation equation has to be effective (based on a 1 cm^3):

Table 8.1 Structural properties of specimens

Wood species	Early wood			Late wood		
	d_i (μm)	n_i (piece/cm²)	a	d_i (μm)	n_i (piece/cm²)	b
Thuja	26.5	142,800	0.8482	14.0	316,600	0.1518
Spruce	30.0	111,335	0.8478	19.0	160,400	0.1522
Pine	28.0	125,100	0.6694	20.0	135,840	0.3306
Larch	38.0	65,490	0.6310	17.5	145,000	0.3690
Beech (vessel)	66.0	15,740	0.7000	48.0	14,020	0.3000
Beech (tracheid)	8.2	342,890		6.4	490,290	
Oak (vessel)	260.0	400	0.5900	35.7	12,000	0.4100
oak (tracheid)	22.5	120,000		19.6	85,000	
b. locust (vessel)	230.0	546	0.5800	120.4	1 500	0.4200
b. locust (tracheid)	15.0	270,000		9.6	280,000	
Cottonwood (vessel)	69.7	9500	0.6666	44.0	12,700	0.3333
Cottonwood (tracheid)	12.7	309,500		11.0	300,892	
Ash (vessel)	177.0	670	0.6100	52.00	750	0.3900
Ash (tracheid)	19.0	190,000		14.0	230,000	

$$\rho_t = \left[\left(1 - \frac{d_1^2 \cdot \pi}{4} \cdot n_1 \right) \cdot a + \left(1 - \frac{d_2^2 \cdot \pi}{4} \cdot n_2 \right) \cdot b \right] \cdot \rho \qquad (8.1)$$

where
ρ_t, ρ bulk density and real density of wood, respectively
d_1, d_2 mean diameter of cavities in the early and late wood
n_1, n_2 number of cavities on the unit surface in the early and late wood
a, b early and late wood portion ($a + b = 1$).

We can use the above equation also to double-check the early and the late wood separately, based on the knowledge that the bulk density of late wood is approximately two times higher than that of early wood.

Table 8.1 summarizes the typical characteristics of the internal structure of the conifer and hardwood species tested.

Cutting vessels, tracheids and other elements of the texture causes surface irregularities. An important basic data is gained by quantifying the number of vessels cut on a certain length in the direction of processing. The scattering of the vessel diameters usually shows normal distribution, which enables the utilization of the mean diameter size without making a greater error.

The position of the vessels measured to the surface is always accidental, which obviously causes a scattering of the surface roughness obtained.

Adding up the number of structure elements cut on the surface gives a characteristic measure of structural surface roughness as shown in the model in Fig. 8.12 (Magoss and Sitkei 2000, 2001).

The area of the valleys has a connection with the number and diameter of structure elements measured on a given unit of length in the processing direction, which is expressed in the following equation:

$$\Delta F = \frac{\pi}{8} \left[a \cdot \left(\sqrt{n_1} \cdot d_1^2 + \sqrt{n_2} \cdot d_2^2 \right) + b \cdot \left(\sqrt{n_3} \cdot d_3^2 + \sqrt{n_4} \cdot d_4^2 \right) \right] (\text{cm}^2/\text{cm})$$

(8.2)

where
n_1, n_2 the number of vessels and tracheids in the early wood in the unit cross-section
n_3, n_4 the number of vessels and tracheids in the late wood in the unit cross-section
d_1-d_4 the diameter size of vessels and tracheids in the early and late wood, respectively
a, b portions of early and late wood

The value ΔF defined with the Eq. (8.2) is called *structure number*, which gives an accurate definition of each wood species based on the size and specific number of cavities in the wood structure. Accordingly, surface irregularities caused by the internal wood structure are expected to have a definite correlation with the structure number.

A further advantage of the structure number is that it enables the characterisation of wood species based on their internal structure, and it helps to establish correlations among the surface roughness parameters.

There is a good correlation between the diameter of vessels or tracheids and their specific number. This correlation for European conifers and hardwood species is depicted in Fig. 8.13. Their distribution is hyperbolic (a straight line on a double logarithmic scale) and conifers have a slightly wider tracheid diameter for the same specific number in the unit cross-section. This diagram is also suitable to check new measurement data.

The curves are well described with the following equation:

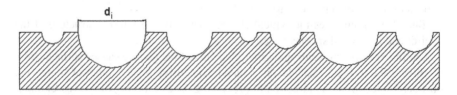

Fig. 8.12 The model of structural surface roughness

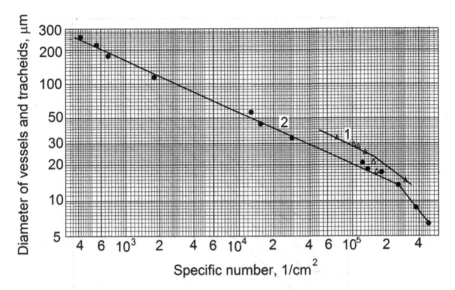

Fig. 8.13 Correlation between the diameter of vessels and tracheids and their specific number in the unit cross-section 1—Conifers; 2—hardwoods

for hard woods $d = 3300 \cdot N^{-0.43}$ [μm]

for conifers $d = 4230 \cdot N^{-0.43}$ [μm]

where N means the specific number of vessels or tracheids.

Using the above equation, the contribution of vessels in the early and latewood can be expressed as follows

$$\Delta F_i = a \cdot 1.574 \cdot d_i^{0.8372} \; (\text{mm}^2/\text{cm})$$

$$\Delta F_i = b \cdot 1.574 \cdot d_i^{0.8372} \; (\text{mm}^2/\text{cm})$$

where a and b mean the portion of early and latewood, and d_i diameter must be substituted in mm. The relationship is demonstrated in Fig. 8.14.

Example: in Oakwood the average vessel diameter is 260 μm in the earlywood and 60 μm in the latewood. The early and latewood portions are $a = 0.59$ and $b = 0.41$. Their expected contributions are $0.3007 + 0.0612 = 0.362$ mm²/cm.

Detailed calculations show that the contribution of tracheids to the structure number is from 10% to 15%. Taking the average value of 10%, oak should have a structure number of $\Delta F = 0.402$ mm²/cm.

The most common values of structure number for selected wood species are summarized in Table 8.2.

It is well-known that results of surface roughness tests usually show significant scatterings. One of the reasons for this is the accidental position of tracheids and vessels to the machined surface. A more significant scattering can be caused by the

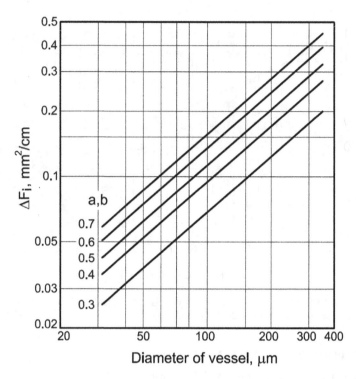

Fig. 8.14 Prediction of structure number for known vessel's diameters in the early and latewood

Table 8.2 Structure number for different wood species

Wood species	ΔF mm^2/cm
Conifers	Around 0.1
Larch	0.11–0.14
Cottonwood	0.15–0.20
Beech	0.19–0.25
Ash	0.25–0.30
Black locust	0.33–0.43
Oak	0.40–0.55

accidental position and cut of the early and late wood or the seasonal change of the early wood/late wood ratio, respectively. Namely the value ΔF may have substantial differences in the early and late wood. Pine species show the smallest divergence, whereas hard wood species generate a significantly bigger one. Figure 8.15 illustrates the alteration of the ΔF value of different wood species depending on the early wood ratio. The starting point of the curves ($a = 0$) indicates the pure late wood, while the end point ($a = 1$) indicates the pure early wood. Oak shows the biggest change, but ash shows a significant one as well (Magoss et al. 2005).

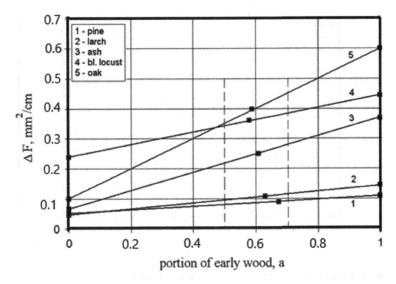

Fig. 8.15 Alteration of the ΔF value of different wood species in relation to the early wood ratio

Tests show that the early wood ratio falls predominantly in the range $a = 0.5$–0.7 as shown in Fig. 8.15.

Therefore, it seemed practical to apply the relative changes of ΔF for the range $a = 0.5$–0.7 in accordance with the following equation:

$$\delta(\Delta F) = \frac{\Delta F_{0.7} - \Delta F_{0.5}}{\Delta F_{0.6}} \tag{8.3}$$

Wood species can be characterized with the ratio of cross sections of cavities cut in the early and late wood, which we can express using the Eq. (8.2) as follows:

$$B = \frac{d_1^2 \cdot \sqrt{n_1} + d_2^2 \cdot \sqrt{n_2}}{d_3^2 \cdot \sqrt{n_3} + d_4^2 \cdot \sqrt{n_4}} \tag{8.4}$$

Since pine species have the vessels missing, thus the Eq. (8.4) for conifers will be simpler.

Figure 8.16 shows the relative changes of ΔF in relation to the parameter B defined with the Eq. (8.4).

The apparent correlation between the two variables is displayed clearly. However, it is remarkable that beech is located right next to Scotch pine; and black locust shows a smaller relative change than larch. Consequently, the relative change has no connection with the absolute value of ΔF. The correlation obeys on the following empiric equation:

$$\delta(\Delta F) = 7.8 \cdot B^{0.65} (\%) \tag{8.5}$$

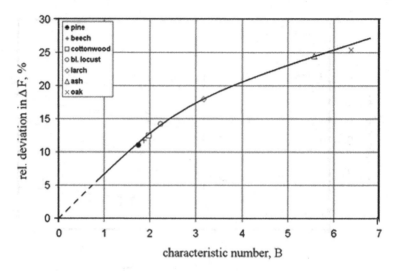

Fig. 8.16 Relative changes of ΔF in relation to the parameter B

8.5   The Origin of Surface Roughness

Roughness that evolves during machining has two major components: machining-caused roughness and roughness caused by the internal wood structure. Even in the case of an ideal machining rough surface evolves due to the inner cavities cut. Moreover in the recent practice of high-speed machining, roughness due to machining is usually much less than the structure-based roughness, especially in the case of hardwood species with large vessels.

The roughness due to machining usually depends on the following factors: cutting speed, chip thickness, processing direction relatively to the grain, rake angle of the tool, sharpness of the tool edge (tool edge radius) and vibration amplitude of the work piece.

Wood cutting belongs to the group of the so-called free cutting. Its main characteristic feature is the absence of a counter-edge, therefore, the counterforce is produced by the strength of wood and forces of inertia. The higher strength and hardness parameters (modulus E) the wood has the smaller force of inertia is required; that is to say, the slower the roughness increases with the decrease of the cutting speed.

The primary reason for machining-caused roughness is the brittle fracture of the material and its low tensile strength perpendicular to the grain. The occurrence of brittle fracture cannot be eliminated, however, it can be limited to a lower volume. The most effective method for this is the high-speed cutting and the smallest material contact possible (sharp tool edge).

The only way to eliminate the negative effect of the tensile strength perpendicular to the grain is to generate a compression load in the immediate vicinity of the edge. This can be facilitated with a 65–70° tool cutting angle or a 20–25° rake angle.

Especially the edge machining of boards is very inclined to breaking the edge because of the tensile load; therefore the selection of appropriate kinematic conditions is very important (Westkämper and Freytag 1991).

Excessive compression load deforming the material can also cause roughness. The method 'smooth machining' has been known for a long time, which is based on the knowledge that smaller chip thickness raises smaller forces. Compression load can also be reduced by using the 'slide cutting'. Slide cutting produces shear load on the edge, which also contributes to material failure. In accordance with the well-known equation the equivalent stress has the following form:

$$\sigma_e = \sqrt{\sigma^2 + 4\tau^2} \tag{8.6}$$

The distribution of compression load inside the material depends on the tool edge radius (bluntness of the tool edge). Due to the radius of the edge, a layer in the material of thickness z_0 will be compressed underneath the edge (Fig. 8.17).

This layers thickness is about 70% of the tool edge radius. The load of the edge is transferred onto the material on a surface with a $2b$ width and a length that is identical with that of the edge. This is similar to the strip load. The biggest load appears just under the contact surface and it rapidly decreases as we move towards the inside of the material. This means that the highest compression of cells always starts directly underneath the edge.

If the compression load underneath the edge reaches the ultimate bearing stress of the material the cell system suffers permanent deformation and it gets compressed to the detriment of the cavities (Fischer und Schuster 1993). If out of total deformation z_0 the permanent deformation is z_1, then the expelled material will be located in

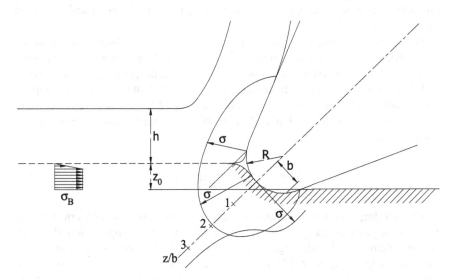

Fig. 8.17 Compression effect of the edge

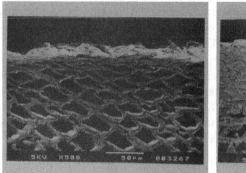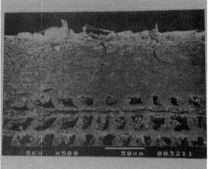

Fig. 8.18 Compaction behaviour of early and late wood layers caused by the same blunt edge

the lower layers and the thickness of the compressed layer z_2 is accordingly (see Figs. 8.17 and 8.18):

$$z_2 = z_1 \cdot \rho_1/(\rho_2 - \rho_1) \qquad (8.7)$$

where

ρ_1 is the volume density without compression
ρ_2 is the volume density of the compressed material (1–1.2 g/cm^3)

The volume density of early and late wood is significantly different, while the value ρ_2 can change only to a limited extent; therefore we can expect the following values after compression: approximately $z_2 = z_1$ in the early wood and $z_2 \cong 3z_1$ in the late wood.

The above statement was also observed experimentally (Fischer und Schuster 1993). Figure 8.18 shows the compaction of early and late wood using the same blunt tool. The late wood was compacted several times deeper than the early wood.

The deformation underneath the edge is elastic in the first period of the compression; it will therefore rebound once the edge has passed by. The approximate rate of elastic deformation can be calculated easily. When the elastic half space is exposed to a strip load:

$$\sigma = \frac{E}{2 \cdot \left(1 - v^2\right)} \cdot \frac{z}{2b} \qquad (8.8)$$

The sanding is an unusual cutting process due to its more or less spherical cutting edge with negative rake angle. An interesting problem is the comparison of the effect of a grit element and knife edge on the deformation of wood structure just under the sliding edge (Magoss 2013). Based on the engineering mechanics (Boussinesq problem), calculations were carried out to compare the stress distributions under knife edge and grit, Fig. 8.19. The cutting edge geometry modify the stress distribution

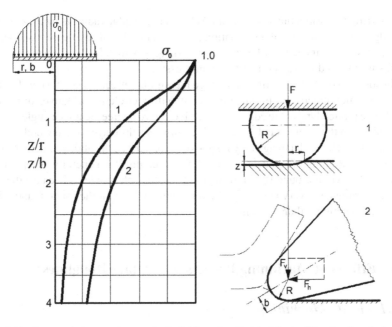

Fig. 8.19 Stress relations under a spherical (1) and cylindrical body (2) using hemispherical pressure distribution on the contact surface

and a bigger radius of a sphere is equivalent to a smaller radius of a cylindrical edge concerning their effect on the deformed layer. This deformed layer fundamentally determines the expected R_k values which is experimentally verified.

If for the purpose of simplicity we presume that the elastic deformation z_r is sustained till the crushing stress σ_B is reached, then from Eq. (8.8) we get:

$$z_r = \frac{\sigma_B 4b\left(1 - \nu^2\right)}{E} \tag{8.9}$$

In the case of pinewood compression load between the directions B and C can have values $\sigma_B = 15$ N/mm^2, then $R = 10$ μm radius is likely to produce elastic deformation of approximately $z_r = 1$ μm, while a $R = 50$ μm radius is expected to bring elastic deformation of approximately $z_r = 5$ μm. These values give one seventh of the total deformation expected ($z_0 = 7$ μm and 35 μm, respectively). The above results would have the logical consequence that the compression of the upper layer should be considered in each case.

The wood structure however contains cavities; and even cavities of smaller size can be measured to the radius of a sharp tool edge. Therefore the sharp tool edge can penetrate the cavities and break the cell walls. The tool edge bends these broken wood parts standing vertically, which then take the necessary deformation z_0 without transferring it towards the lower layers. When the tool edge radius is increased, both the edge size and the rate of deformation z_0 exceed the size of cavities; consequently, a compression of the surface layer evolves. An approximately $R = 20$ μm tool edge radius is expected to trigger a surface compression.

In some respect, the jointing of an edge has similar effects as edge wear. Jointing is the common practice to produce the same cutting circle for all knives of a cutter head. The jointed land at the edge, however, has a zero degree clearance angle, which, depending on the jointed land width, will compress a thin layer of the wood causing cell damage. This cell damage may responsible for surface instability and, in the case of subsequent gluing, for a decrease in gluing strength. It is generally accepted that narrow land width around 0.25 mm will not create such problems because the area of zero clearance is very small. It is also recommended that the jointed land width would not be higher than 1.0 mm.

8.6 Effects of Machining Process on Surface Roughness

8.6.1 Knife Machining

Machining operations may have effects on the surface roughness of any wood material. The resulting surface roughness is influenced by an array of operational parameters which have quite different importance. The main operational parameters are the following:

- cutting speed,
- cutting depth,
- feed per tooth,
- cutting angle,
- workpiece vibration,
- sharpness of tools,
- climb or counter cutting,
- oblique cutting,
- cutting circle radius,
- direction of cut.

The tool interacts with the wood material and its mechanical properties influence the surface roughness. The following material properties should be mentioned:

- volume density,
- hardness of wood,
- modulus of elasticity,
- anisotropy of wood,
- early and latewood,
- inhomogeneity of wood, including irregular growth and defects.

Due to the "free cutting" mechanism, the most important influence factor is the cutting speed (Dobler 1972). The counter force to the cutting edge is ensured by the strength of the material and by inertia forces. Soft woods have less strength and therefore they are more sensitive to the selection of a proper cutting speed. This is clearly seen in Fig. 8.20.

Increasing the cutting speed, the average height R_z and the valley depth R_{vk} steeply decrease, while the core depth R_k slightly decreases.

Beech wood has more strength and therefore the inertia forces are less important. The increase of cutting speed moderately decreases the R_z and R_{vk} values and the core depth R_k is rather constant, Fig. 8.21. Around 30 m/s the counter force is well developed and increasing the cutting speed does not lower the roughness much.

The valley depth R_{vk} proportionally increases with the vessel diameter. Therefore, the *Abbott-ratio* (AR) defined as $(R_{pk} + R_k)/R_{vk}$ is a further characteristic number in a dimensionless form which is distinct for each wood species. The Abbott-ratio is a sensitive parameter with a wide variation. Its maximum value is around 1.6 for conifers, while its lowest value is around 0.15. It correlates well with the structure number ΔF as depicted in Fig. 8.22.

The relationship between Abbott-ratio and structure number is given by the equation

$$\frac{R_{pk} + R_k}{R_{vk}} = \frac{0.35}{\Delta F^{0.76}} - 0.4 \qquad (8.10)$$

or

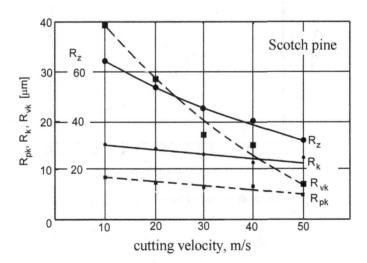

Fig. 8.20 Effects of cutting speed on the surface roughness parameters in the case of Scotch pine

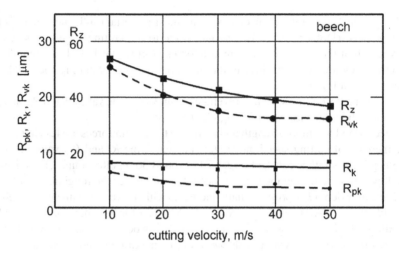

Fig. 8.21 Effects of cutting speed on the surface roughness parameters in the case of beech

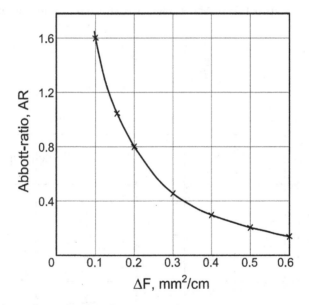

Fig. 8.22 Abbott-ratio as a function of structure number

$$\Delta F = \left(\frac{0.35}{\frac{R_{pk}+R_k}{R_{vk}} + 04} \right)^{1.32}$$

    The introduced Abbott-ratio sensitively responds to the behaviour of different wood species for varying cutting speeds. Using the measured data in Figs. 8.20 and

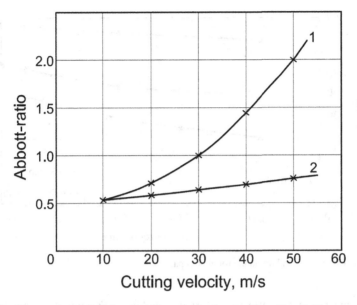

Fig. 8.23 Abbott-ratio as a function of cutting velocity for pine (1) and beech (2)

8.21, the calculated Abbott-ratios are depicted in Fig. 8.23. At low cutting velocity (10 m/s), the Abbott-ratio is the same for both species, but the reason for it is not the same. Scotch pine suffers higher deformation and brittle fracture under the cutting edge resulting in large valley depth R_{vk}. Beech wood has smaller R_{vk} value in spite of bigger cavities compared to scotch pine. Scotch pine responds to higher cutting speeds with steeply decreasing R_{vk} values, which strongly increases the Abbott-ratio.

The tooth bite has also some influence on the surface roughness. It is well known that the essence of a smooth machining is to use a small feed per tooth as the last operation. A small tooth bite means less cutting force which causes less deformation and break-out in the deformation zone.

Figure 8.24 shows experimental results for three wood species as a function of tooth bite (Magoss 2008). The variation of R_z for all three wood species is quite similar and the curves can be described with the empirical equation

$$R_z = A + B \cdot e_z^n \tag{8.11}$$

where e_z is the tooth bite in mm. The exponent n for all species may be taken as 0.6, the constant B is also the same for all species with the value of 43 (R_z is given in μm). The meaning of the constant A is interesting and it is very close to the minimum roughness of each wood species corresponding to its apparent anatomical roughness in Fig. 8.50. The best fit for A provides the values of 69, 35 and 19 for oak, beech and Scotch pine respectively. This finding is important and allows an approximate estimation of roughness for other wood species.

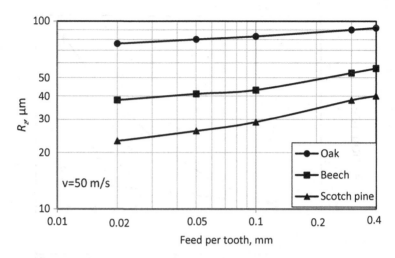

Fig. 8.24 Effect of tooth bite on the roughness parameter R_z

Workpiece vibration affects the surface roughness. Vacuum clamping on CNC-machines allows only small-amplitude vibrations of the workpiece, not exceeding 5–10 μm amplitude. Our measurements have shown that no difference in the roughness was observed between vacuum and rigid clamping of the workpiece. If the tooth bite is higher than 1 mm, then the increased vibration amplitude may worsen the surface roughness due to uneven tooth bite..

Other conditions prevail when the workpiece moves during the woodworking operation. In this case the vibration amplitude may be much higher, perceptibly worsening the surface quality (Sitkei and Gyurácz 1990).

The vibration of the workpiece mainly depends on the following factors:

- the size of the workpiece (thickness, length),
- the magnitude of exciting forces and their direction,
- the instantaneous boundary conditions, i.e. the position of pressing rolls, their pre-stressing and spring stiffness characteristics (Sitkei et al. 1985, 1988, see Sect. 5.4).

The magnitude of exciting forces depends on the sharpness of the tool and the tooth bite. If the exciting force acts towards the machine table, the vibration amplitudes are smaller, producing less roughness. On the contrary, exciting the workpiece towards the pressing rolls, the vibration amplitudes may be high enough to considerably worsen the surface quality.

A special vibration mode is "wash-boarding", mainly occurring with bandsaw and circular saws. Wash-boarding is a wavelike surface pattern appearing on the surface of sawn timber, Fig. 5.25 in Chap. 5. The profile can be as much as 0.6–0.8 mm deep on the face of the lumber requiring larger target size in order to remove sawing deviations and to reduce the depth of washboard (Okay et al. 1995; Lehmann and Hutton 1997; Orlowski and Wasielewski 2001).

Wash-boarding is a self-excited resonance phenomenon occurring with high blade speed, feed speed and thinner saw blades. The frequency of the tooth passage should be slightly higher than the natural frequency of the blade (Okay et al. 1995). Wash-boarding occurs in a narrow frequency range and it can be eliminated by increasing or decreasing the blade speed, or the bandsaw tension.

The cutting angle also influences surface roughness. Higher cutting angles around 65–70° exert more compressive forces into the material ensuring a smoother surface. On the contrary, the use of a small cutting angle in cutting veneer requires a pressure beam to provide the necessary compressive stresses (Sitkei and Gyurácz 1990, see Sect. 1.5).

There are advantages to oblique cutting, which utilizes the combine effect of compressive and shear stresses. This cutting requires less compressive forces and produces less deformation and a better surface quality (Sitkei 1997). For the same reason very thin veneers can only be cut by using large oblique angles. The action of pure shear stresses is utilized when cutting upholstering (foam) materials.

8.6.2 Sanding

Sanding is important among the different woodworking operations and it is an abrasive process. Due to the unusual cutting edge, with a negative rake angle and the random position of grits on the surface, it can produce a smooth surface, depending on the grit size and other operational parameters (Scholz and Ratnasingam 2005; Siklinka and Ockajova 2001).

The grit size of sandpaper is determined by the number of meshes per inch of the sieves used for screening. The average grit diameter is depicted in Fig. 8.25 as a function of standard grit size notation.

Sanding is a very common woodworking operation to produce smooth surfaces or to adjust a given thickness (e.g. particle board). It is widely accepted that surface sanding proved to be the most advantageous processing step prior to coating and painting (Richter et al. 1995 and see Sect. 8.8.9).

Due to its almost spherical cutting edge, sanding is an unusual cutting process with some distinct differences compared to knife cutting. The main differences are caused by the compaction and clogging effects of grits always exerting compressive forces onto the material.

In Sect. 8.4 the mechanics of the sanding process was treated in detail and a comparison between knife cuttings and sanding was also presented. An important finding was that a knife with a given edge radius compacts the underlying layer twice as deeply as grit does with the same radius. Another important conclusion is that the compaction exerted by the edges is directly related to the core depth R_k.

The most important operational parameters of sanding are the cutting speed, surface pressure, grit size and feed speed. They determine the specific wood removal $(g/m^2 \cdot min)$ and the surface roughness parameters for a given wood species. It is well known that the specific wood removal continuously decreases as a function

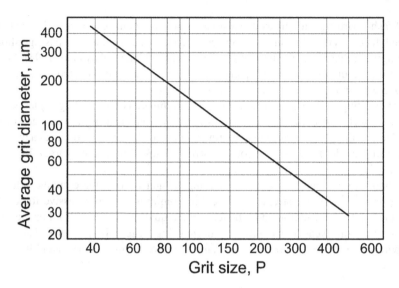

Fig. 8.25 Average grit diameter as a function of standard grit size

of working time due to the wear process. The linearity between stock removal and surface pressure in a wider range does not hold. At higher pressures more and more chips are in the contact zone hindering the penetration of grits into the material and resulting in a decrease of wood removal. Similarly, wood removal is theoretically independent of grit size. However using fine grits, wood removal may decrease considerably. Readers interested in the relationships describing the interaction of the operational parameters are referred to Sect. 3.8.

In the following the results of a large-scale experimental work will be presented to explain the basic regularities of the sanding process. Five different wood species (spruce, larch, beech, black locust and oak) were selected with known structural properties. In these experiments four grit sizes (P-80, P-120, P-150 and P-240) were used with sanding velocities around 25 m/s and with surface pressure between 0.25 and 0.35 N/cm^2. Ten measurements were performed on each sample using a perthometer with a tip radius of 2 μm.

A detailed roughness representation for beech wood is given in Fig. 8.26 as a function of average grit diameter. The relationship is almost linear for all roughness components. The strong dependence of the R_k-layer is striking. This finding refers to the strong influence of the particle diameter on the crushing depth in the surface layer.

The smoothness of a surface is considerably determined by the reduced peak heights. This relationship for spruce and beech is given in Fig. 8.27 on a double logarithmic scale. The decrease of grit size creates quite low values which would give a good polishing possibility using wax or light-colored oils. The polishing ability of a surface plays a definite role in the colour enhance of a given wood species (Sitkei 2013).

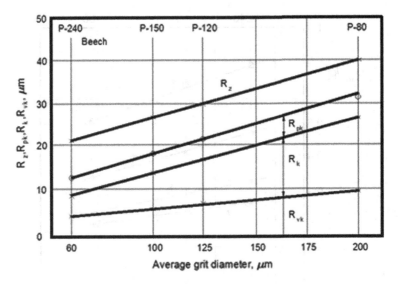

Fig. 8.26 Effects of grit size on the surface roughness parameters for beech wood

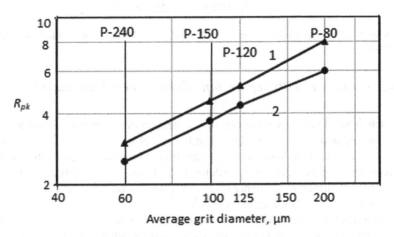

Fig. 8.27 Relationship between the reduced peak heights R_{pk} and grit size 1—Spruce, 2—Beech

A further evidence for the crushing effect of grits is shown in Fig. 8.28 (Csanády et al. 2015). This Figure shows the R_{vk}/R_z ratio for five wood species as a function of grit diameter. A larger grit causes more deformation which reduces the valley depth. Big vessel species suffer more clogging and more reduction in the valley depth.

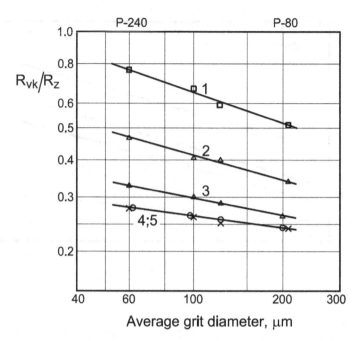

Fig. 8.28 The R_{vk}/R_z ratio as a function of grit size 1—Oak, 2—Black locust, 3—Larch, 4—Spruce, 5—Beech

8.6.3 Problems with Separation of Roughness Components

A much debated question is the possible separation of roughness components due to machining and anatomy, especially for big vessel wood species. Tracheids have dimensions in the same range as roughness irregularities do. Their contribution to the resultant roughness can hardly be determined. Furthermore, tracheids will fundamentally be damaged in the upper layer of the new surface and, as a consequence, they are not visible in the roughness profile.

Any attempt in this direction will be conditional until, as long as the roughness of perfectly cut samples is not known. To our best knowledge, such measurements are not published yet. But in the possession of such results there will be further difficulties. Woodworking operations, both knife machining and abrasive sanding, cause deformations modifying the structure of upper layer significantly. Measuring the surface roughness, we scan the irregularities due to machining and the irregularities of a modified bottom layer. Especially tracheids and smaller vessels are prone to be untraceable. Therefore, looking at a roughness profile, we cannot identify any tracheids, only the bigger vessels.

The above difficulties are strengthened by some measurement results obtained with Japanese planer. Japanese planer is able to cut extreme clean surface the tracheids of which are becoming distinct in the roughness profile (see Fig. 8.10). To our surprize, the measured roughness values, especially Ra, were rather higher than lower compared to machined surfaces (planing, milling, sanding) for the same wood species. This result might have been interpreted such that the machining does not cause any additional roughness or even reduces the anatomical roughness.

In reality, the measured roughness is a composition of irregularities due to machining and those of randomly scanned structural elements modified by the edge, and bigger vessels. Therefore, if we suppose that the resultant roughness is the sum of the components due to machining and wood anatomy then we should concern an *apparent anatomical roughness*. Latter is not identical with that of the perfectly cut surface because it was deformed and modified by the cutting edge.

Big vessels on hardwood surfaces are fully visible in the roughness profile and several attempts were made to separate their contribution to the resultant roughness. This attempt is justified by the fact that the contribution of vessels may be much higher, than the contribution of machining and in this case the judgement of machining is more difficult. We are of opinion, however, that we may have more information with counting than omitting something. Namely, without exception, everybody wanted to eliminate vessels from the roughness profile using some kind of filtering method (Fujiwara et al. 2003; Gurau et al. 2005; Csiha and Krisch 2000).

An other similar problem was arisen when evaluating bigger surface joined from several boards. In this case grain run-out may change from place to place causing quite different number of vessels in the trace length (Csanády et al. 2019). It is obvious that the measured roughness also varies in an unexpected extent. In order to solve the problem, different approaches may be followed.

The simple use of the number of cut vessels in the trace length though shows the tendency of roughness variation but with considerable scatter, Fig. 8.29.

A much more accurate method is to measure the area of the cut vessels in the trace length and to plot the roughness parameter as a function of total area of cut vessels. These measurements are shown in Fig. 8.30 for milled oak. This representation shows a close correlation and it is also suitable to read the intersection ($R_a = 2 \ \mu m$) for zero cut vessels. This value is comparable or even better than values for conifers.

The linear relationship has the simple form:

$$R_a = 2 + 116.8 \cdot A_n$$

where A_n means the total area of cut vessels in the roughness profile in mm^2.

Keeping in mind Fig. 8.15, an interesting comparison may be done. An oak has an average vessel cross-section of 0.4 mm^2/cm (ΔF values) but the maximum measured value in Fig. 8.30 is only 0.14 mm^2/1.25 cm. This finding is explained by the fact that the Perthometer cannot measure the true profile of cut vessels. The average roughness is integrated from the measured profile which is the reason for the strong correlation in Fig. 8.30.

Finally, it may be concluded that neither the roughness component due to machining nor the component due to internal structure can accurately be determined. They are strong intermingled between themselves, except bigger vessels. Making some assumptions, an apparent roughness due to machining can be established (see Sect. 8.8).

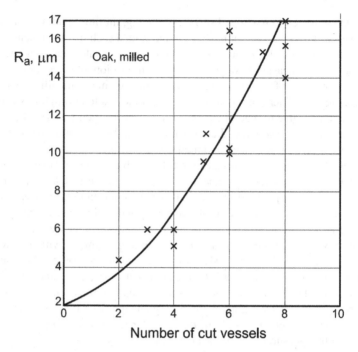

Fig. 8.29 Average roughness R_a as a function of number of cut vessels in the trace length

8.6.4   Effect of Tool Wear on the Surface Roughness

It is a well-known fact that enhanced wear of the tool increases the surface roughness. This is the ultimate practical reason why tools are re-edged based on a certain working time or cutting length (see in Chap. 7).

Blunt tools with a bigger edge radius transmit bigger forces on the material; the material in front of the tool travels a longer distance going around the edge. The forces transmitted on the chip at the detachment point cause a fracture of elementary particles. Fractures beneath the cutting level are primarily expressed in the Abbott parameter R_k; therefore, this parameter is expected to be highly sensitive to tool edge deterioration.

Figure 8.31 shows the changes of the R_k parameter of four different wood species when using two different tool edge radii.

It is clearly visible that the parameter R_k showed a twofold increase in each case in comparison to cutting with sharp edges. These data lead us to conclude that the parameter R_k gives a good feedback on the deterioration status of the tool edge.

The evidence of destroying effect of a blunt tool can be detected in several ways. Many cases the surface roughness profile clearly shows such phenomena. As an example, Fig. 8.32 shows roughness profiles for sharp and blunt tools.

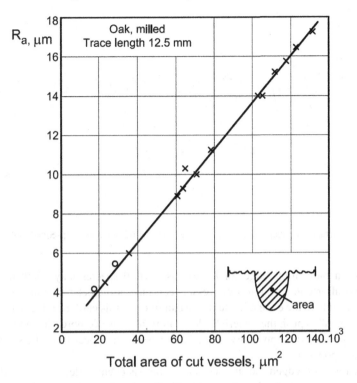

Fig. 8.30 Average roughness R_a as a function of total area of cut vessels

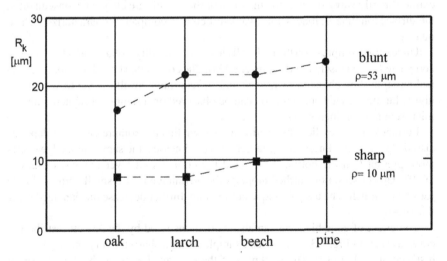

Fig. 8.31 Changes of the R_k value when using sharp and blunt tool edges in relation to four different wood species

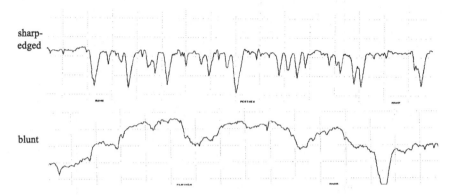

sharp-
edged

blunt

Fig. 8.32 Surface roughness profiles of oak machined with blunt and sharp-edged tools

Using a blunt tools edge for large-vessel wood species may result in surface waviness after compression. Oak vessels can have a diameter size up to 250–300 μm: a size where the edge of a blunt tool can fit in. In this case the edge will not only crumple but also push the material. These motions cause compression and waviness in the upper layer; the majority of the large vessels disappear from the surface due to the compression.

When a sharp-edged tool is used, the surface is even; valleys are caused by the vessels and tracheids cut. Using a blunt tool ($R = 53$ μm), at the same times, gives an extraordinarily wavy surface, the majority of the vessels are clogged. Consequently, the surface roughness alone is not always sufficient to completely characterize surface quality.

Under high compression the cell walls lose their stability, collapse and form a fully compacted layer (Fischer und Schuster 1993). The compacted layer has poor mechanical strength, low abrasion resistance and suffers quick swelling due to wetting or higher humidity. This phenomenon can be observed on a sample of Scotch pine at a microscopic magnification, Fig. 8.33.

Furthermore, an earlier observation has shown that the compressed layer depends also on the chip thickness: using larger chip thicknesses, the same blunt edge exerts more pressure on the bottom layers and the compressed layer increases (Sanyev 1980). This observation further supports the advantage to use small chip thickness for surface finish. The use of oblique cutting may further decrease the damaged layer thickness.

The effect of edge blunting can also be demonstrated by the Abbott distribution curve as seen in Fig. 8.34 for a beech sample using a sharp tool (1) and after a feed distance of 1800 m (2). The thickness of the deformed zone doubled as the sharp edge blunted due to the feed distance.

Investigating the effect of edge blunting on roughness parameters, the tool edge radius or the feed distance can be chosen as an independent variable. The primary variable is the edge radius, which directly influences the expected surface roughness.

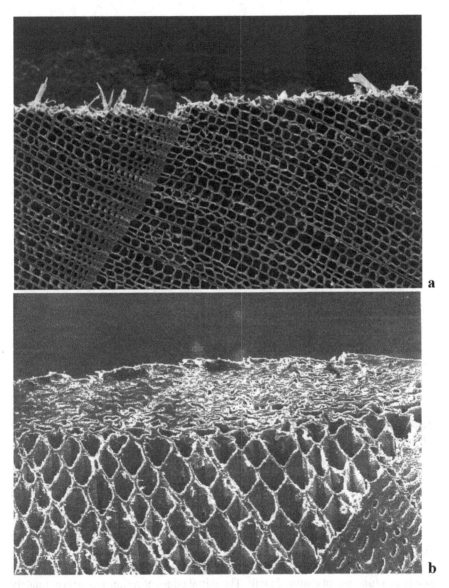

Fig. 8.33 Clear cut of a sharp tool edge (**a**) and compression of the surface as a result of a blunt tool edge (**b**) (Fischer and Schuster 1993)

We distinguish the total feed distance and the true cutting length during the lifetime of a tool.

Keeping in mind Sect. 8.7.3 and especially Eqs. (7.3), (7.4) and (7.5), the total cutting length can be calculated as

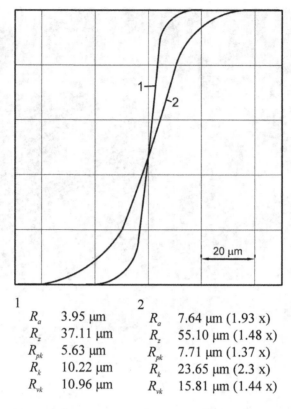

1		2	
R_a	3.95 µm	R_a	7.64 µm (1.93 x)
R_z	37.11 µm	R_z	55.10 µm (1.48 x)
R_{pk}	5.63 µm	R_{pk}	7.71 µm (1.37 x)
R_k	10.22 µm	R_k	23.65 µm (2.3 x)
R_{vk}	10.96 µm	R_{vk}	15.81 µm (1.44 x)

Fig. 8.34 Abbott distribution curves for beech wood milled with a sharp tool (1) and after a feed distance of 1800 m (2)

$$L_c = \left[\frac{1}{A \cdot v^n} \cdot \left(\frac{\rho}{\frac{\sin(\beta/2)}{1-\sin(\beta/2)}} - y_0 \right) \right]^{1/m} \tag{8.12}$$

Knowing the total cutting length, from Eq. (7.4) the lifetime of the tool can be estimated.

Figure 8.35 shows measurement results for particleboard with two different cutting speeds as a function of cutting length. The initial edge recession was approximately 20 µm which corresponds to an initial edge radius of approximately 17 µm. The exponents in Eq. (7.3) are $n = 0.78$ and $m = 0.65$, and the constant is $A = 0.0235$. Edge radius and edge recession are given in µm.

The tool edge radius usually increases the roughness parameter R_z nearly in a linear way. Figure 8.36 illustrates the correlation that was established by testing Scotch pine and beech samples.

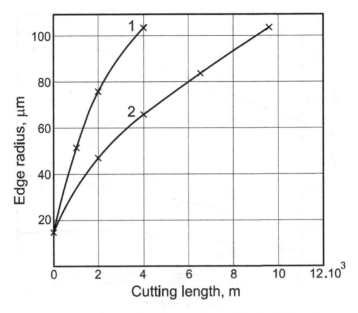

Fig. 8.35 Edge radius as a function of cutting length at milling Particleboard, cutting velocity 40 m/s (1) and 20 m/s (2), tungsten carbide

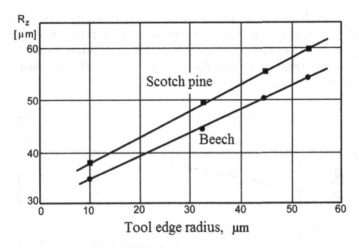

Fig. 8.36 Correlation between the parameter R_z and the tool edge radius (Magoss and Sitkei 2001)

While the edge wear has a parabolic character as a function of cutting length (Fig. 8.35), the different roughness parameters follow a similar fashion. Figure 8.37 represents different roughness parameters with increasing feed distance. The core depth R_k is more than doubled strongly indicated edge wear.

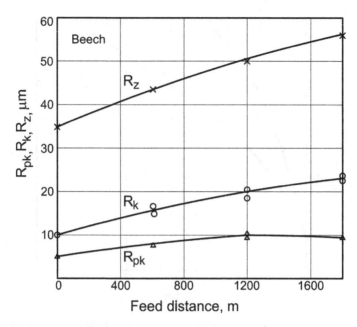

Fig. 8.37 Variation of roughness parameters as a function of feed distance

Figure 8.38 shows characteristic ratios for the same beech wood samples from Fig. 8.37. The increase of the R_k/R_z ratio indicates that the core depth R_k grows more rapidly than R_z as a function of working time. The same is true for the R_a/R_z ratio. Although R_a is an integrated value, it is more responsive to structural changes in the surface layer due to edge wear.

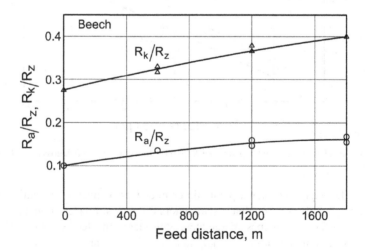

Fig. 8.38 Variation of roughness ratios as a function of feed distance

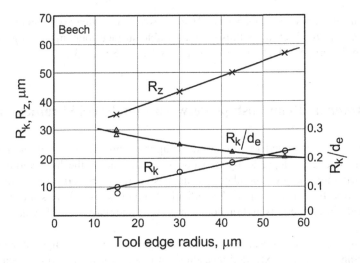

Fig. 8.39 Effect of tool wears on the roughness parameters R_z and R_k

The most sensitive and useful roughness parameter is the core depth R_k that uniquely indicates the change in the edge radius due to the wear process. The roughness parameter R_k, is almost constant (around 10 μm) for a wide variety of wood species using sharp tools. These initial values will be at least doubled as shown in Fig. 8.37.

Quite similar results were obtained for other wood species (conifers, Black locust, Oak). Therefore, these results may be regarded as generally valid.

The reduced peak height R_{pk} also monotonously increases with advancing wear. To produce a smooth surface, this may be a limiting factor concerning the allowable edge wear. Choosing the edge radius as the influencing variable, the roughness parameters R_z and R_k vary linearly as a function of the edge radius as depicted in Fig. 8.39.

In the figure the R_k/d_e ratio (d_e is the equivalent edge diameter) is also given which characterizes the effect of compaction by the edge in the core zone. In sanding this ratio is nearly constant for all wood species, see Fig. 8.57.

Using knife edges, the same ratio is higher and slightly increases toward sharp edges. This may be explained by the general theory of wood cutting (Sitkei and Gyurácz 1990). The radial component of the cutting force for sharp edges changes its sign from negative to positive. Due to tension forces in the separation zone, some fibres may be torn out from the developing surface and slightly increases the core depth. With a higher edge diameter, the radial force is always compressive which does not create any additional roughness.

Tool wear is a major factor influencing the surface quality obtainable by any machining. Furthermore, tool wear increases energy requirements. Therefore, the maintenance of tools is very important and the sharpening cycle of tools is determined by surface quality requirements. If tools are sharpened more frequently than necessary, the higher maintenance costs will lower the economy of production.

As outlined above, the tool wear and surface roughness intensively interact. Therefore, not only the lifetime but also the allowable surface roughness is a governing factor in the proper selection of tool materials (see in Sect. 7.3).

8.7   Internal Relationships Between Roughness Parameters

To generalize our knowledge in surface roughness problems, functional relationships are needed independently of wood species. Due to the high variability of wood structure, the great number of roughness parameters and possible measurement errors, this task seems to be difficult. Roughness parameters are imperfect statistical representations and random sampling is one reason that they are prone to scatter.

Noting these difficulties, it is useful to search possible relationships among roughness parameters using extended measurement results for many wood species. The functional variables may be single, combined or even dimensionless quantities. The obtained relationships may have different accuracy mainly depending on the relative scattering of measurement data and the proper selection of procession technique.

The average roughness R_a and the irregularity depth R_z are linearly correlated to each other having a given scattering zone, Fig. 8.40 (Csanády et al. 2015; Magoss 2008). Their ratio varies between 0.09 and 0.12, and the bigger ratios are generally from wood species with wider voids (vessels). It is known that R_a depends on the

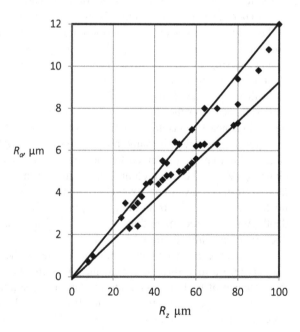

Fig. 8.40 Diagram of R_a versus R_z for different wood species

cross-section of voids, while R_z depends on the depth of irregularities which explains the inherent scattering of experimental results.

Measurement results on samples surfaced with the Japanese planer have further revealed that a more perfect cut without surface deformation allows to scan all voids which increase the integrated average roughness R_a and the R_a/R_z ratio may be as high as 0.17.

At the same time, their correlation is strong enough for practical uses. It is worth to mention that the average roughness R_a has a better correlation with the sum of Abbott parameters and their correlation obeys the following empirical equation

$$R_a = 0.275 \cdot \left(R_{pk} + R_k + R_{vk} \right)^{0.85} \tag{8.13}$$

which is depicted in Fig. 8.41.

The average roughness height R_z has a similar correlation function which is given by

$$R_z = 2.41 \cdot \left(R_{pk} + R_k + R_{vk} \right)^{0.85} \tag{8.14}$$

which corresponds to the averaged straight line in Fig. 8.40.

Due to the definition of S_z in the 3D system, it does not correspond to R_z and similar relationship for S_z cannot be established. At the same time, S_a and Abbott parameters are fully useful. Their correlation described by the equation

$$S_a = 0.268 \cdot \left(S_{pk} + S_k + S_{vk} \right)^{0.85} \tag{8.13a}$$

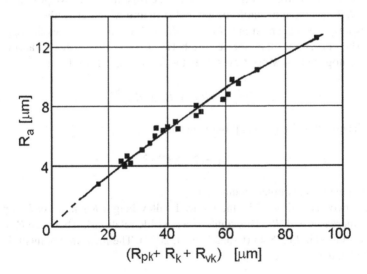

Fig. 8.41 Relationship between R_a and the *Abbott*-parameters (Magoss and Sitkei 2001)

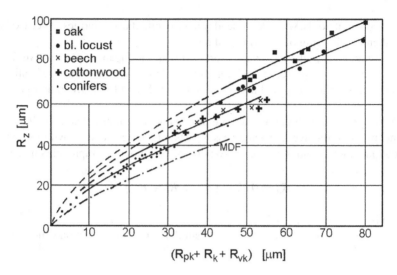

Fig. 8.42 Relationship between irregularity depth R_z and the *Abbott*-parameters

which agrees well with the corresponding R_a function in Fig. 8.41.

As it is seen in Fig. 8.40 between R_a and R_z there is only a loose interrelation. As a consequence, no uniquely defined relationship between R_z and the sum of Abbott parameters can be expected. Nevertheless, the experimental results depicted in Fig. 8.42 show an interesting picture.

A lot of curves are obtained. For a more accurate explanation, the measurement results on MDF-boards of different volume density included (Magoss and Sitkei 2000). MDF has the most uniform internal structure which gives the lowest curve. Oak has large vessels and a much less uniform structure and gives the uppermost curve. The curves for other species are between these two extremes according to their inhomogeneities. Most curves obey the following general form

$$R_z = A \cdot \left(R_{pk} + R_k + R_{vk}\right)^{0.65} \tag{8.15}$$

where the constant A can be approximated as

$$A = 7 \cdot \Delta F^{0.225}$$

and ΔF must be substituted in mm^2/cm.

In hardwoods with vessels, the reduced valley height R_{vk} more or less proportionally increases with R_z. This suggests a possible relationship between R_a and R_{vk} which is plotted in Fig. 8.43 (Csanády et al. 2019). The measured points follow the empirical equation

$$R_a = 0.41 \cdot R_{vk}^{0.82}$$

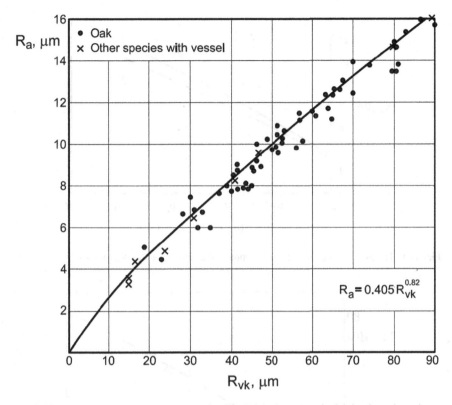

Fig. 8.43 Average roughness versus reduced valley height for oak and other hardwood species

with acceptable accuracy. This relationship is only valid for hardwood species with vessels. The abrasive sanding operation exerts more crushing effect onto the surface depending on the grit size. The sum of Abbott parameters is, however, more or less insensible to these crushing effects and the average roughness R_a has similar relationship to that of Eq. (8.13) for knife machining. The grit size influences the average roughness only in a smaller extent, Fig. 8.44. Using Eq. (8.13), the constant has values of 0.175 and 0.248 for P-80 and P-240 respectively.

Similarly, the roughness height parameter R_z has very good correlation with the sum of Abbott parameters independently of grit size and for the tested five wood species. Figure 8.45. The curve is described by the following equation

$$R_z = 3.065 \cdot \left(R_{pk} + R_k + R_{vk}\right)^{0.75} \mu m$$

The exponent is slightly lower compared to the R_a function, similarly to Fig. 8.42. Wood species: Spruce, larch, beech, black locust, oak.

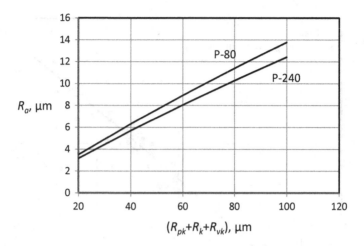

Fig. 8.44 The average roughness R_a as function of the sum $(R_{pk} + R_k + R_{vk})$ for sanding

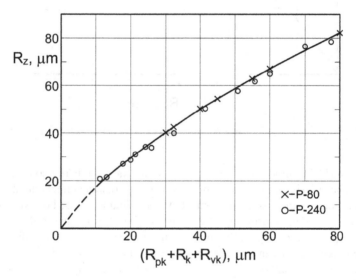

Fig. 8.45 Relationship between roughness height R_z and the sum of Abbott parameters. Cutting speed 25 m/s, platen pressure 0.3 N/cm^2, grit sizes between P-80 and P-240.

8.8 Skewness and Kurtosis

These surface roughness parameters are defined and described in Sect. 8.8.2. In the past, the wood industry did not commonly use height distribution parameters of *skewness* and *kurtosis*. They deserve more attention because they can provide useful information about the topography of a wood surface. The negative skewness means that the peak of the height distribution is located near the surface. A kurtosis higher than 3 means a more spiked distribution.

Results with different wood species are summarized in Fig. 8.46. There is a close correlation between skewness and kurtosis, independent of wood species, which obeys a quadratic equation with the form

$$R_{ku} = 3 + 1.48 \cdot R_{sk}^2$$

Conifers have little skewness and kurtosis and that means that the height distribution nearly corresponds to their normal distribution. With increasing hardness, both skewness and kurtosis will increase considerably.

The skewness and kurtosis relationship is strongly connected to the R_q/R_a ratio. As skewness and kurtosis increase, the corresponding R_q/R_a ratio also increases. A detailed analysis of the R_q/R_a ratio (Csanády et al. 2015), shows a close connection to kurtosis: for more spiked height distributions the R_q/R_a ratio uniquely increases. On a very dense (1.2 g/cm^3) rosewood (Dalbergia cochinchinensis), $R_q/R_a = 2.4$, $R_{sk} = -6$ and $R_{ku} = 45.7$ values were measured. Figure 8.46 shows that there is more kurtosis for the same skewness when the R_q/R_a ratio increases. This phenomenon was mainly observed in conifers.

The correlation between skewness and R_q/R_a ratio is depicted in Fig. 8.47. Conifers generally have R_q/R_a ratios between 1.2 and 1.5, mostly grouping around 1.3. The R_q/R_a ratio slightly increases when using a small tooth bite of 0.1 mm or less. Hardwoods may have quite high R_q/R_a ratios up to 2.4 with a corresponding high negative skewness. The correlation can be approximated by a linear function which intersect on the horizontal axis

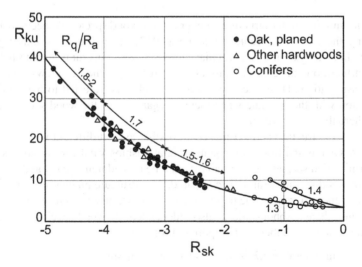

Fig. 8.46 Interrelation between skewness and kurtosis for different wood species

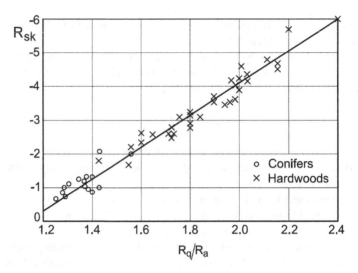

Fig. 8.47 Correlation between skewness and R_q/R_a ratio

$$-R_{sk} = 4.69 \cdot \left(R_q/R_a - 1.12\right)$$

although the curve theoretically tends to the zero skewness if R_q/R_a tends to one.

8.9   The Use of Structure Number and Abbott-Ratio

The determination of the structure number ΔF for wood species has become feasible by using the data from Table 8.1 and Eq. (8.2). The structure number gives a unique characterization of a particular wood species concerning its internal structure. Furthermore, differences caused by the area where the tree grew will be taken into consideration. Therefore, the structure number is expected to have a definite correlation with the roughness parameters, regardless of the wood species and the area where they were grown.

The use of structure number makes it possible to treat different wood species in a unified system and to discover general relationships valid for all wood species. The term "wood species" as a variable cannot be treated numerically and, therefore, different wood species cannot be compared. The structure number uses anatomical properties and differences due to growing conditions can be taken account.

The introduced Abbott-ratio is also independent of wood species but it differs from the structure number in an important property:

- the structure number ΔF is independent of machining,
- the Abbott-ratio AR is dependent of machining,
- ΔF and AR are interrelated only for similar machining conditions.

In order to prove the usefulness of structure number, ten European wood species, soft- and hardwoods, were selected. Samples were processed on a special super surfacer (so-called horizontal surfacer), where the R_Z component of roughness caused by machining did not exceed 10 μm. The relationship established by evaluating 10 different wood species is shown in Fig. 8.48 (Magoss and Sitkei 2000).

This curve demonstrates the least surface roughness that can be achieved in practice as a function of the structure number. The relationship can be described with the following empiric equation:

$$R_z = 122 \cdot \Delta F^{0.55} \tag{8.16}$$

In Sect. 8.5.3 we have discussed the inherent difficulties encountered with the separation of roughness components. In order to have some insight into this problem, the following approximation and approach have been used. To separate the roughness components, three 20 by 5 cm samples were tangentially cut from each wood species. After machining they were finished by a special finishing machine. The finishing was repeated until the profile was flat and suitable for evaluation. Establishing the finished surfaces, the same samples were subjected to milling operations using various cutting speeds between 10 and 50 m/s. These surfaces were evaluated with the common surface measuring methods.

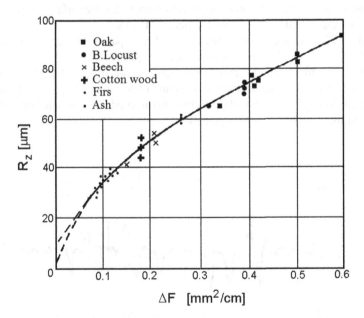

Fig. 8.48 Relationship between irregularity depth R_z and the structure number ΔF based on the parameters of 10 wood species

A hypothetical base line was first established on the finished surfaces. The corresponding R_z'-value was calculated taking only the positive amplitudes into consideration (Fig. 8.49). This is the roughness component due to woodworking operations. Knowing the overall R_z value and subtracting the roughness component due to woodworking, we get to an R_z value due to the internal structure of wood.

Identification the machining roughness enables us to calculate and plot an apparent anatomical roughness which would be caused by the anatomy alone.

Figure 8.50 shows the apparent anatomical roughness and the actual roughness as a function of structure number at two different cutting speeds.

A general relationship for Fig. 8.50 can be expressed in the following empirical form:

$$R_z = \left(123\Delta F^{0.75} + 35e_z^{0.6}\right) \cdot \left(1 + \frac{50 - v_x}{50} \frac{0.1183}{\Delta F^{0.83}}\right) \qquad (8.17)$$

$$10\,\text{m/s} \le v_x \le 50\,\text{m/s}$$

where ΔF must be substituted in mm^2/cm, e_z in mm, and v_x in m/s. The third part of Eq. (8.17) as well as the curves clearly illustrate that the softer pine wood is more sensitive to a decrease of the cutting speed. This phenomenon can be explained by the lesser local rigidity of pine.

Using a sufficiently high cutting speed, it appeared that the surface roughness will mainly be determined by the internal structure of wood.

In the following we discuss surface roughness parameter ratios which show uniquely defined correlations with the structure number ΔF.

Figure 8.51 illustrates the correlation between the relationship R_a/R_k and the structure number ΔF. The anatomical structure of wood causes a fivefold variation in the R_a/R_k ratios. This leads to the conclusion that wood species cannot be compared on the simple basis of surface roughness.

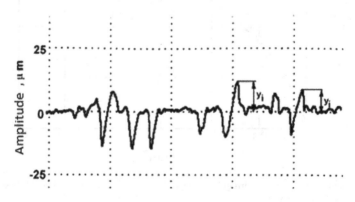

Fig. 8.49 Measurement of the roughness component due to machining (Magoss and Sitkei 2003)

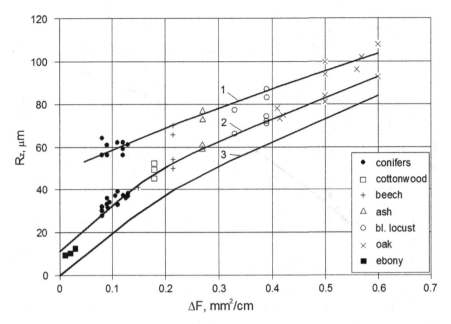

Fig. 8.50 Irregularity depth R_z, related to different cutting speeds in relation to the structure number ΔF 1-cutting speed 10 m/s; 2- cutting speed 50 m/s; 3-apparent anatomical roughness (Magoss and Sitkei 2003)

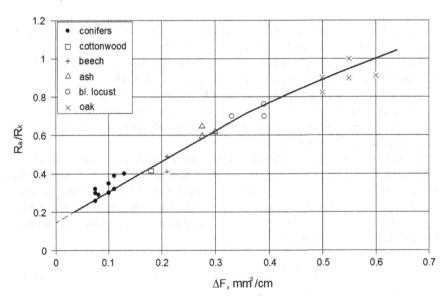

Fig. 8.51 Correlation between the relationship R_a/R_k and the structure number ΔF (Magoss and Sitkei 2003)

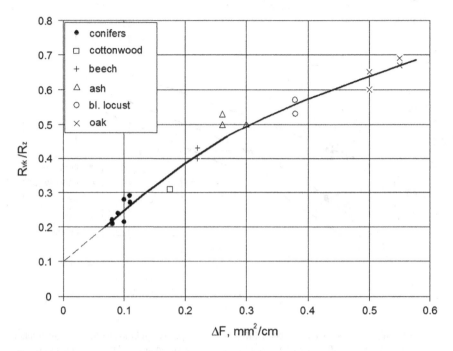

Fig. 8.52 Correlation between the relationship R_{vk}/R_z and the structure number ΔF

Figure 8.52 shows the correlation between the structure number ΔF and the R_{vk}/R_z ratio. Here we have a threefold increase in this ratio.

Finally we examined how the core depth of the material ratio curve (R_k) affects the surface roughness as a function of the structure number ΔF.

Figure 8.53 demonstrates that the value of R_k influences the roughness to a greater extent in soft wood species. It should be noted that the correlation curve in Fig. 8.53 is valid for sharp tools only. The value of R_k is dependent on tool sharpness for all wood species (see Sect. 8.5.4).

As outlined in the introduction of this Section, the Abbott-ratio may vary due to machining. First of all a change in the core depth R_k will shift the Abbott-ratio for the same wood species. Figure 8.54 demonstrates the effect of grit size on the variation of AR (Csanády et al. 2019). A higher grit size increases the core depth shifting the curve upwards. The effect of grit size obvious and its influence is taken into account by the following approximation.

$$\frac{R_{pk} + R_k}{R_{vk}} = 11.42 \cdot d^{0.4} \cdot (1 - \Delta F^{0.2}) \qquad (8.18)$$

where the average grit diameter d must be substituted in mm. The clogging effect diminishing the apparent structure number can be seen in the Figure.

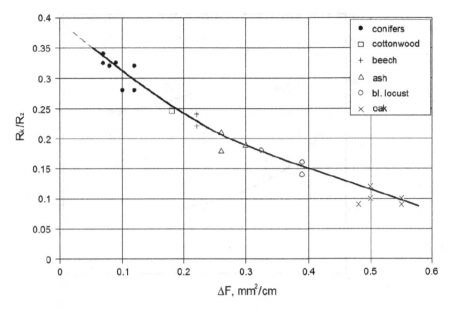

Fig. 8.53 Correlation between the relationship R_k/R_z and the structure number ΔF (Magoss and Sitkei 2003)

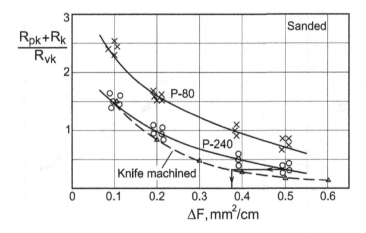

Fig. 8.54 Effect of sanding on the Abbott ratio

At the same time, different wood species can be plotted as a function of Abbott-ratio as it is seen in Fig. 8.55. The roughness height R_z is strongly dependent on the grit size for all species (cutting speed is 20 m/s, platen pressure is 0.3 N/cm^2).

The Abbott-ratio provides a more accurate description of the internal structure of wood in the immediate surface layer taking into account the modifying effect of the woodworking operation. Figure 8.56 shows the relative valley depth as a function of

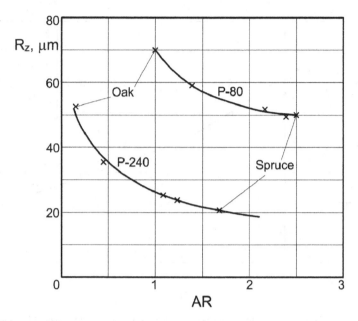

Fig. 8.55 Average roughness R_z as a function of Abbott-ratio for sanded surfaces. Wood species: Spruce, larch, beech, black locust, oak

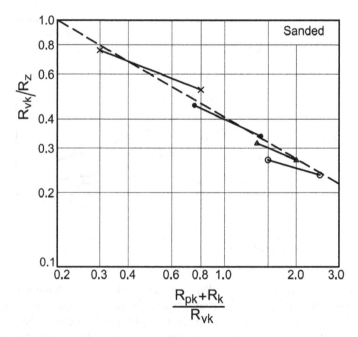

Fig. 8.56 The relative valley depth as a function of Abbott-ratio

Abbott ratio on double logarithmic scale using the results of Fig. 8.26 (Sect. 8.5.2). The results are well approximated with the following simple equation

$$\frac{R_{vk}}{R_z} = \frac{0.406}{\left(\frac{R_{pk}+R_k}{R_{vk}}\right)^{0.56}} \tag{8.19}$$

These results uniquely demonstrate the usefulness of the Abbott-ratio in seeking general relationships which highly facilitate the processing of new experimental results or the approximate estimation of different roughness parameters (Csanády et al. 2019).

A further detailed analysis of obtained experimental results revealed that the ratio of core depth R_k and the average grit diameter is practically constant and it does not depend on the grit size and wood species as shown in Fig. 8.57 (Magoss 2013). Its ratio is averaged to 0.1 which makes it possible to forecast the expected core depth R_k in advance or to select an appropriate grit size to achieve a given R_k layer thickness.

Clogging decreases the number and size of cavities in the surface. As a consequence, some roughness parameters such as R_z may be smaller than the expected anatomical roughness value. This is more explicitly demonstrated in Fig. 8.58 which gives an overview on the variation of R_z values as a function of the structure number ΔF (Magoss 2013).

Conifers are not prone to clogging. Over $\Delta F = 0.2$ (beech) the clogging effect appears and remains for all vessel-species although its extent decreases for big-vessel species. Clogging can also be characterized by the Abbott-ratio, as demonstrated in Fig. 8.54.

It should again be stressed that big vessel species, such as oak, may have higher scattering of roughness parameters concerning an extended surface. The number of

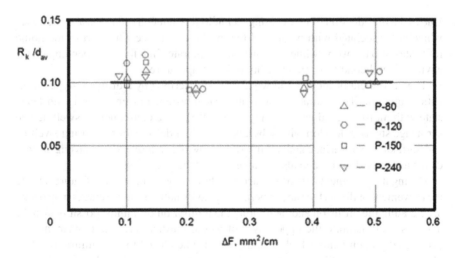

Fig. 8.57 Core depths and grit diameter ratio depending on the structure number ΔF.

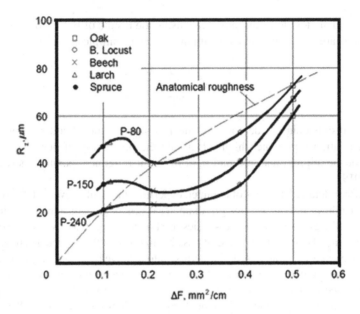

Fig. 8.58 An overview of R_z values for wood species as a function of structure number ΔF

cut vessels in the trace length influences the measured value, also at sanding, in considerable extent.

8.10   Influence of Wetting on Surface Roughness

Due to the introduction of new coating technologies, aqueous coatings are often used which are associated with wetting of the machined surface. One part of the liquid infiltrates into the wood while the other part evaporates into the air depending on the environmental conditions (relative humidity, temperature).

The movement of moisture in wood can be described by a diffusion-type differential equation. The characteristic feature of moisture movement is its unsteady state with moisture gradient; (Csnády et al. 2015). As a consequence, swelling and shrinkage stresses develop with subsequent residual deformations. During swelling and shrinkage, the surface and a thin layer below are in movement, causing surface distortion (raised grain) and also changes in surface roughness.

During its movement a wood surface may have another interesting feature. Due to the movement of the surface and to permanent deformations, micro cracks may occur, accelerating the diffusion and facilitating the infiltration of liquids, possibly also the aqueous coatings, into the upper layer of wood. Furthermore, the movement may occasionally open voids which would otherwise be closed to free infiltration. This phenomenon may influence not only the surface roughness but also the adhesion properties of coating films on wood surfaces depending on the surfacing process (planing or sanding etc.) and wood anatomy (Richter et al. 1995; Magoss 2008, 2013).

Wetting properties of a wood surface fundamentally depend on the anatomical structure and surface deformations caused by machining. Therefore, the infiltration velocity is in strong interaction with the surface integrity.

No systematic research has been found on the subject and therefore, a detailed experimental research program was undertaken to clarify the fundamental relationships concerning different wood species (spruce, Scotch pine, larch, oak, black locust, beech and poplar) and surfacing techniques (planing, sanding, milling, finishing and thermo-smoothing). Distilled water was used for wetting and the initial thickness of the film was around 100 and 180 μm. In order to follow the changes in the surface just after the wetting, a contact-less 3D laser measuring unit was used. All important 3D parameters, including the amplitude parameters such as S_q (RMS), skewness and kurtosis of surface height distribution were recorded and processed (Molnár 2018; Csanády et al. 2015).

In order to get reliable average values, measurements were made on 10 different spots of each sample one after the other at one minute intervals. The measurement started five minutes after wetting and therefore, average values were given only after 15 min. Generally, after one hour already no changes were registered, although the measurements were continued for two hours.

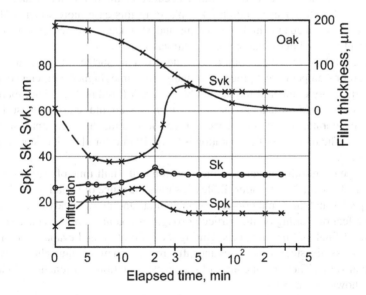

Fig. 8.59 Variation of roughness parameters during the wetting process. Oak planed. Initial film thickness 180 μm

The typical course of a wetting process for planed oak sample with large vessels is illustrated in Fig. 8.59 showing the momentary values of S_{pk}, S_k and S_{vk} roughness parameters.

After 5 min of infiltration the record began with significant deviations compared to the initial values. The R_{pk} and R_k values increased during the infiltration time. On the contrary, the R_{vk} value rapidly decreased, (with all probability it is due to swelling), to a minimum one and after some 15 min rapidly increased to values higher than the initial one. All three parameters have a local maximum culminating in subsequent elapsed times according to their vertical positions. The local maximum of the lower S_{vk} layer coincides with the half time of moisture evaporation (30 min). After the local maximum point, all three parameters proceed asymptotically to their final values which are higher than their initial values. After one hour no change was observed.

The strict variation of all three roughness parameters indicates intensive movements and deformations in the upper surface layer with permanent deformations.

The anatomical structure of wood is a major influencing factor in the simultaneous infiltration-evaporation process and determines the proportion of liquid infiltrated into the wood. The machining deforms the upper thin layer which generally enhances the rate of infiltration. Because the depth of infiltration is higher than the deformed layer, the effect of machining does not surpass that of the anatomical structure.

The rate of evaporation is the highest from a free liquid surface above of which the vapour is saturated. If rapid infiltration occurs then the free water surface rapidly vanishes and a slower evaporation takes place. In this case more water infiltrates to deeper layers causing higher swelling and shrinkage stresses which develop subsequent residual deformations and additional roughness.

Figure 8.60 shows the evaporation of water from the surface of spruce and black locust at room temperature, Spruce has roughly three times higher coefficient of moisture conductivity compared to oak and black locust (Shubin 1973) and, therefore, most of the water infiltrates into the wood which later should return to the surface in order to evaporate. As a consequence, spruce shows a much lower apparent evaporation rate. The half time of evaporation is 9 and 19 min for black locust and spruce respectively.

The type of machining causes some variation in the half time of evaporation and the retained water after one hour. Table 8.2 shows the range of variation for different wood species and machining operations including sanding (Csanády et al. 2015).

The effect of wetting on the particular roughness parameters is characterized by the ratio of final and initial values. During the infiltration and evaporation process the roughness first steeply increases and, after reaching a maximum value, converges to its final value. The typical course of roughness variation as a function of elapsed time is shown in Fig. 8.61.

The experimental results have uniquely shown that the sanded surfaces had always higher variations compared to planed surfaces. That means that a sanded surface can more easily be wetted. Experience has shown its advantage for coating a surface (Richter et al. 1995). Concerning the Abbott-parameters, the reduced peak height shows the highest variation.

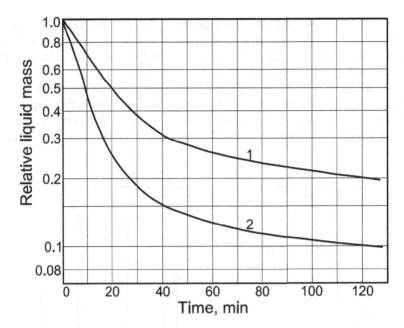

Fig. 8.60 Evaporation of water as a function of time with initial water thickness of 100 μm. 1—spruce, 2—black locust

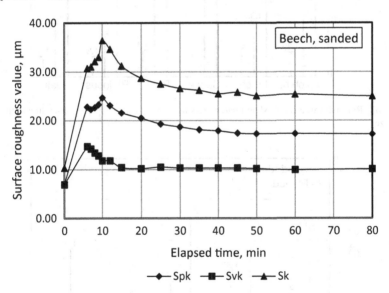

Fig. 8.61 Variation of momentary roughness parameters after wetting as a function of time. Beech sample, sanded. Initial film thickness is 100 μm

Figure 8.62 shows comparative measurements for different wood species when sanding with P120 grit size. The highest variation was always observed for the reduced peak height S_{pk}, especially with big vessel species. The reduced valley height R_{vk} always shows the smallest relative increase.

The core depth S_k (or R_k) plays the most important role in the wetting process. Its relative change in the wetting process depends on the kind of machining. Figure 8.63 represents a comparison of changes in the S_k layer thickness for planing and sanding using different wood species. Sanding always produces a higher relative increase of S_k.

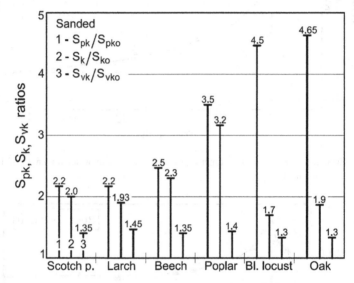

Fig. 8.62 Relative variation of Abbott parameters for different wood species at sanding with P120 grit size

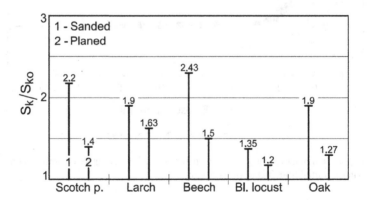

Fig. 8.63 Comparison of wetting properties of different wood species using planed and sanded surfaces

For every thickness of a water film, the evaporation rate influences the amount of water infiltrating into the surface layer. Drying at room temperature, the halftime of evaporation fluctuates between 18 and 20 min. Using forced convection with hot air between 50 and 60 °C, the halftime of evaporation was reduced to 8 min which allowed to much less water penetrate into the deeper layers.

Figure 8.64 shows comparative measurements for evaporation at room temperature and hot air drying. In the latter case the relative increase of roughness parameters is considerably less than with the slow evaporation due to the higher infiltration time.

In conclusion, the wetting of a surface has a definite influence on the surface properties and its roughness after wetting may considerably differ from its initial value. This is especially true for a sanded surface which produces a better surface for coating.

The effect of machining on the wetting properties of wood surfaces can also be measured with the following simple method. Dropping water onto the surface, the initial contact angle decreases continuously due to spreading and infiltration in the first seconds and later infiltration only. Because these phenomena are exponential, the plot of the relative contact angle as a function of time on a half-logaritmic scale yields two straight lines, Fig. 8.65. It is clearly seen that the infiltration on sanded surface is much faster and the straight lines in both phases are considerably steeper. The initial contact angles on the two surfaces differ not much and, therefore, they are less suitable indicator compared to the course of infiltration (Csanády et al. 2019).

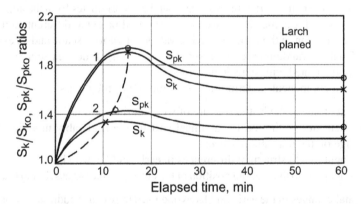

Fig. 8.64 Effect of surface evaporation rate on the relative increase of roughness components. Initial water film thickness is 100 μm, 1—drying at room temperature, 2—drying with hot air (50–60 °C)

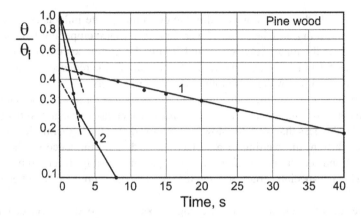

Fig. 8.65 Relative decrease of contact angles as a function of time on Pine wood. 1—planed, Θ_i = 73°, 2—sanded with P-80, $\Theta_i = 70°$

8.11 Scattering of Roughness Data

The measured roughness profile contains partly irregularities due to machining and partly the modified structural irregularities accessible for scanning all of which are of random nature.

In general, the major part of the resultant roughness originates from the anatomical structure of wood. Cavities in the wood material will be cut at different angles and positions in relation to the surface, leaving valleys of various shape and sizes on the surface. That is a random process causing the scattering of the measurement results and it may be treated with statistical methods.

The main components determining the scattering of roughness are

- component due to the brittle fracture of wood during cutting;
- component due to the local anatomical structure considerably modified by the tool edge on the traceable surface;
- component due to structural differences in early and latewood;
- component due to the random position of the cutting plane to vessels and tracheids.

Seasonal changes in the early and latewood ratio may cause additional scattering. The high variability of structural properties within a given species further complicates the problem. The most important relationships of structural variations were treated in Sect. 8.3.

Machining also contributes to the scattering of roughness. The primary reason for the roughness caused by machining is the brittle fracture of wood cells and its low tensile strength perpendicular to the grain. The brittle fracture cannot be eliminated, but it can be limited to a low volume. The most effective method is high speed cutting with the smallest material contact possible (using a sharp edge).

To eliminate the negative effect of low tensile strength perpendicular to the grain, it is important to generate a compressive load in the immediate vicinity of the cutting edge. This can be done using cutting angles in the range of 65–70° or rake angles of 20–25°. The edge machining of boards requires special attention because they are very inclined to the edge breaking due to tensile loads. In order to avoid edge breaking, the selection of appropriate kinematic conditions is very important (see Sect. 8.5.1).

An excessive compression load crushing the material can also cause additional roughness. The method of "smooth machining" has been known for a long time and it is based on the experience that a small chip thickness raises smaller forces. Using oblique cutting, the compressive loads will be considerably reduced due to the contribution of shear stresses in the material fracture (Sitkei 1997).

Data scattering as a random process can be handled with statistical methods. It is necessary to have a set of data which correspond to normal distribution for reliable statistical evaluations. Using a probability net, the plot of experimental results gives a straight line provided that the set of data fulfils the above requirement. At least 60 to 100 measurement data are required to achieve this aim.

Figure 8.66 shows typical measurement results on milled samples of Scotch pine, ash and oak (Magoss 2008). The curves are straight with some deviations at their ends due to their incomplete distribution which has a simple physical explanation: infinitely low and high values do not occur in real distributions. Using the figure, the

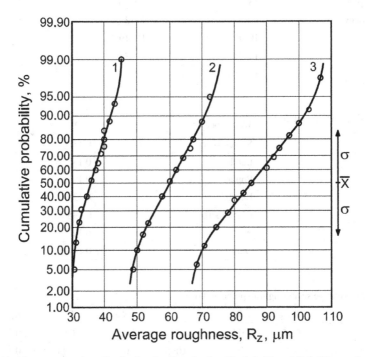

Fig. 8.66 Measurement data distribution for Scotch pine (1), Ash (2) and Oak (3) samples

appropriate values of the average roughness height and its standard deviation can be read. The standard deviation related to the average value gives the relative scattering which is summarized in Table 8.3 for different wood species.

The relative scattering of the different wood species falls in a surprisingly narrow range, between 0.12 and 0.14. This circumstance facilitates the rough estimation of the scattering based on the average values. Further experiments have revealed that the different roughness parameters may have dissimilar relative scattering. Performing 30 measurements on samples side by side showed relative scattering given in Table 8.4. In this experiments Spruce systematically had lower scattering compared to Oak. R_{max} and R_{pk} are the most sensitive to scattering, while R_a and R_k are less prone to scattering (Table 8.5).

As it is outlined in Sect. 8.5.3. big vessels are less modified by woodworking operations and their number in the measured trace length can considerably influence the roughness. Especially on bigger surfaces, where the run-out angle of vessels to the surface may vary, a bigger scattering of roughness should be reckoned with. In

Table 8.3 Half time of evaporation for different wood species

Wood species	$t_{0.5}$, min	Water after one hour, %
Spruce	16–19	20–26
Larch	13–16	18–22
Scotch pine	13–20	20–26
Black locust	9–13	9–13
Oak	10–12	16–19
Beech	11–15	19–24
Aspen	17–20	20–27

Table 8.4 Relative scattering of R_z values

Wood species	$\sigma/\overline{X}$
Scotch pine	0.13
Larch	0.14
Poplar	0.125
Beech	0.12
Black locust	0.135
Ash	0.14
Oak	0.14

Table 8.5 Relative scattering for spruce and oak

Wood species	R_{max}	R_z	R_a	R_{pk}	R_k	R_{vk}
Spruce	0.15	0.09	0.08	0.11	0.08	0.1
Oak	0.2	0.1	0.095	0.2	0.11	0.15

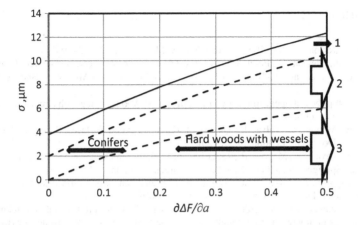

Fig. 8.67 Standard deviation of the roughness parameter R_z as a function of the derivative $\partial \Delta F / \partial a$ for different wood species. 1—roughness component due to machining, 2—component due to random position of cutting plane to vessels and tracheids, 3—component due to structural differences in early and latewood

this case we may have more information using the proposed new processing method compared to the filtering method.

Keeping in mind Fig. 8.15 in Sect. 8.3, the standard deviation referred to R_z can be represented as a function of the derivative in respect to the earlywood portion "a" of the structure number. This relationship, using the data for different wood species, has been constructed formerly (Magoss 2008) including all possible components listed above, Fig. 8.67.

Conifers fall in a narrow range and their standard deviations vary between 4 and 6 μm. Beech and poplar is close to conifers with a range between 5.5 and 6.5 μm. The hardwoods with large vessels occupy a broader range between 8 and 12 μm. All these figures refer to the averaged roughness height R_z.

The measurement of roughness parameters is always associated with a considerable scattering inherent mainly in the internal structure of wood. It should also be added that the measurements are not free from errors. Roughness parameters are imperfect statistical representations due to the very complicated geometrical structures and therefore, they are more prone to scatter as a consequence of random sampling (Thomas 1981).

The scattering of measurement results is the consequence of random processes in the classical sense and it characterizes the random variation of surface topography over a given surface area. Systematic error of measurement may also occur if time to time the measuring unit is not controlled for accuracy (calibration) or some phenomenon disturbs the measurement process. The latter may be typical for optical measurement unit due to the different reflection and colour properties of wood materials (see Fig. 8.3).

Literature

Abbott, E., Firestone, F.: Specifying surface quality. Mech Eng 569–572 (1933)

Butterfield, B., Meylan, B., Peszlen, I.: A fatest háromdimenziós szerkezete—three dimensional structure of wood. English - Hungarian Edition. Faipari Tudományos Alapítvány. Budapest, p. 148 (1997)

Санев, В., Обработка древесины круглими пилами (Woodworking with circular saws). Изд. Лесная Пром. (1980)

Csanády, E., Magoss, E., Tolvaj, L.: Quality of Machined Wood Surfaces, Springer, Heidelberg, New York, Dordrecht, London, p. 257 (2015)

Csanády, E., Kovács, Z., Magoss, E., Ratnasingam, J.: Optimum Design and Manufacture of Wood Products. Springer, Heidelberg, New York, Dordrecht, London, p. 421 (2019)

Csiha, Cs., Krisch, J.: Vessel filtration—a method for analysing wood surface roughness of large porous species. Drevarsky Vyskum **45**(1), 13–22 (2000)

Csiha, Cs., Faanyagok felületi érdességének vizsgálata „P" és „R" profilon, különös tekintettel a nagyedényes fafajokra [Surface Roughness of Wood Using „P" and „R" Profiles for Big-Vessel Hardwood Species], PhD Diss., Sopron (2003)

Devantier, B.: Prüfmethode zur objektiven Bewertung der Rauhigkeit und Welligkeit von Holzwerkstoffen. Abschlussbericht IHD Dresden (1997)

Dobler, K.: Der freie Schnitt beim Mähen von Halmgut. Hohenheimer Arbeiten No. 62 (1972)

Dong, W., et al.: Comprehensive study of parameters characterising three-dimensional surface topography. Wear 29–43 (1994)

Fischer, R., Schuster, C.: Zur Qualitätsentstehung spanend erzeugter Holzoberflächen. Mitteilung aus Institut für Holztechnik der TU Dresden (1993)

Fujiwara, Y., Fujii, Y., Okumura, S.: Effect of removal of deep valleys on the evalution of machined surfaces of wood. Forest Prod. J. 58–62 (2003)

Gurau, L., Mansfield-Williams, H., Irle, M.: Processing roughness of sanded wood surfaces. Holz Roh Werkst. **2005**(63), 43–52 (2005)

Шубин.Г., Физические основы и расчет процессов сушки древесины. (Physical background of wood drying.) Лесная промышленностъ, Москва (1973)

Lehmann, B., Hutton, S.: The kinematics of wasboarding of bandsaws and circular saws. In: Proceedings of 13th IWMS Vancouver, pp. 205–216 (1997)

Magoss, E.: General regularities of the surface roughness sanding solid woods. Proceedings of the 21th International Wood Machining Seminar Tokyo, Japán: Japan Wood Research Society, pp. 325–332 (2013)

Magoss, E., et al.: New approaches in the wood surface roughness evaluation. In: Proceedings of the 17th IWMS Rosenheim, pp. 251–257 (2005)

Magoss, E., Sitkei, G., Lang, M.: Allgemeine Zusammenhänge für die Rauheit von bearbeiteten Holzoberflächen für Möbel. Möbeltage in Dresden, S. 273–279 (2004)

Magoss, E., Sitkei, G.: Optimum surface roughness of solid woods affected by internal structure and woodworking operations. In: Proceedings of the 16th International Wood Machining Seminar, Matsue, pp. 366–371 (2003)

Magoss, E., Sitkei, G.: Fundamental relationships of wood surface roughness at milling operations. In: Proceedings of the 15th International Wood Machining Seminar, pp. 437–446 (2001)

Magoss, E., Sitkei, G.: Strukturbedingte Rauheit von mechanisch bearbeiteten Holzoberflächen. Möbeltage in Dresden, Tagungsbericht S. 231–239 (2000)

Magoss, E., General Regularities of the Wood Surface Roughness. Progress Report No.1, University of Sopron, p. 38 (2008)

Molnár, Zs.: Végmegmunkált természetes faanyagok felületi stabilitása nedvesítéskor (Stability of machined wood surfaces at wetting) PhD Dissertation, Sopron (2018)

Okay, R., et al.: What is relationship between tooth passage frequency and natural frequency of the Bandsaw when Wasboarding induced. In: Proceedings of 12th IWMS Kyoto, pp. 267–380 (1995)

Orlowski, K., Wasielewski, R.: Washboarding during cutting on frame sawing machines. In: Proceedings of 15th IWMS Los Angeles, 219–228 (2001)

Richter, K., et al.: The effect of surface roughness on the performance of finishes. Forest Prod. J. 91–97 (1995)

Saljé, E., Drückhammer, J.: Qualitatskontrolle bei der Kantenbearbeitung. Holz als Roh- und Werkstoff. S. 187–192 (1984)

Schadoffsky, O.: Objektive Verfahren zur Beurteilung der Oberflächenqualität. Tagungsbericht Bielefeld (1996)

Scholz, F., Ratnasingam, J.: Optimization of sanding process. In: Proceedings of 17th IWMS Rosenheim, pp. 422–429 (2005)

Siklinka, M., Ockajova, A.: The study of selected parameters in wood sanding. In: Proceedings of 15th International Wood Machining Seminar Los Angeles, pp. 485–490 (2001)

Sitkei, Gy., Gyurácz, S., Horváth, M.: Zusammenhang zwischen Werkstück Schwingung und Oberflächen Güte bei der Holzbearbeitung. Acta Facultatis Ligniensis, Sopron 1985/1 S. 5–15

Sitkei, Gy., Gyurácz, S., Horváth, M.: Das Schwingungsverhalten des Holzwerkstückes während des Bearbeitung. Acta Facultatis Ligniensis, Sopron 1988/1 S. 5–13

Sitkei, G., Gyurácz, S., Horváth, M.: Theorie des Spanens von Holz, Fortschrittberichte No. 1, Acta Facultatis Ligniensis Sopron (1990)

Sitkei, G.: On the mechanics of oblique cutting of wood. In: Proceedings of 13th IWMS Conference, pp 469–476 (1997)

Sitkei, G., Zur Beurteilung der Farbtöne von Holzarten, Fortschrittbericht No. 2, DWE Sopron (2013)

Thomas, T.: Characterization of surface roughness. Precision Engineering 97–103 (1981)

Westkämper, E., Hofmeister, H.W., Frank, H.: Messtechnisches Erfassen und Bewerten von Massivholzoberflächen. Abschlussbericht AiF Projekt 9681, Braunschweig (1996)

Westkämper, E., Freytag, J.: PKD-Schneidstoff zum Sägen melaminharzbeschicteter Spanplatten. IDR 1, 46–49 (1991)

Chapter 9
Optimization of Wood Machining

9.1 Introduction

Since ancient time Nature has the capability to organize its phenomena optimally. Mankind has also striven for long times to use natural or artificial process to achieve an optimum or the best outcome possible. In the last centuries mankind has developed scientific, mostly mathematical methods for seeking optimal solutions based on different criteria. Basic requirement for any optimization is a well-defined objective function and its constraint functions. In general, we are interested in optimum solution which would fulfil several criteria. For example, we are striving for optimum production cost at the shortest production time possible. In this case a multicriteria optimization is encountered with multiple objective functions and constraints. Because the optimum for several objective functions are in conflict, the overall optimum is obtained either by the use of utility function or compromise.

The objective and constraint functions can correctly be formulated with the use of functional relationships which describe the dependence of output parameters as a function of influencing variables. Such functional relationships are, for example, the operational parameters of various woodworking machines, energy requirements, tool life relationships, surface roughness etc.

In this Chapter the strict mathematical method and an engineering method are described. Several worked examples demonstrate the use of optimization on the field of wood machining.

9.2 General Remarks

The manufacturer of wood products is interested to carry out the necessary manufacturing operations (machining, surface treatment, assembly etc.) to minimize the total manufacturing time and cost, and maximize the quality of the product. This activity

E. Csanády and E. Magoss, *Mechanics of Wood Machining*, https://doi.org/10.1007/978-3-030-51481-5_9

requires careful and many sided decision making based on a properly selected optimization to select the best value of influencing factors giving the highest possible product utility and aesthetical value at a minimum total manufacturing cost.

The first step of this activity is to delineate the problem boundaries of the manufacturing system, the available machines and their capabilities (rotation speed, spindle power, feed speeds, machining accuracy), machining costs and other machine-related costs, quality requirements (surface roughness, tolerance requirements). An objective function, a quantitative criterion should be selected and defined which determines the optimum selection of system variables. In order to accomplish the objective function with the necessary constraints, the functional relationships are needed which express the system behaviour as a function of system variables (Csanády et al. 2019). Essentially this activity formulates and draws up a system model for the solution of the given optimization problem. It should be stressed that the success of an optimization process fundamentally depends on the available functional relationships and their reliability. Generally, if functional relationships are available, then finding the optimum solution does not mean mathematical problems, even in the non-linear cases. Therefore it should be attached great importance to develope the widest circle of functional relationships (Csanády et al. 2019).

Considering machining optimization, it may be treated as a subsystem of the whole production system. A smaller subsystem can be solved individually with less difficulties and the possible interactions among subsystems can be taken into account.

The objective function is a mathematical formulation of selected criterion on the basis of which the performance of the manufacture can be evaluated. Many times an economic criterion is selected such as the minimum production casts, maximum production rate, minimum material use, minimum energy consumption, optimum use of machine capacity.

The different criteria are generally in conflict with each other. It is not possible to simultaneously maximize or minimize several criteria. In a simplified procedure one main criterion can be selected as primary and the others as secondary. The secondary criteria will be treated as constraints with prescribed maximum or minimum values. Therefore decision making is always associated with compromise in selection of primary and secondary criteria according to their importance.

Finally it worth to mention that the idea of optimisation in science is several hundreds of years old. Already Lagrange formulated his famous "Optimal Column" problem around 1770 (Csanády et al. 2019) and the multicriteria optimization has been proposed around 1900 by Pareto (Stadler 1988).

9.3 Scientific Methods of Optimization

The main task of an optimization is to find solution of the objective function subjected to constrains in equality and inequality form. A single objective function with m variables and constraints can be written in the following form:

$$\text{maximize (minimize) } F(x_1, x_2, \ldots x_m)$$

subjected to

$$g_1(x_1, x_2, \ldots x_m) = 0 \quad h_1(x_1, x_2, \ldots x_m) \leq 0$$
$$g_2(x_1, x_2, \ldots x_m) = 0 \quad h_2(x_1, x_2, \ldots x_m) \leq 0$$

$$g_m(x_1, x_2, \ldots x_m) = 0 \quad h_m(x_1, x_2, \ldots x_m) \leq 0$$

In the practice, the above equations are non-linear and analytical solution can not be expected. There are numerical methods, however, such as the Fletcher-Powell algorithm which is widely used successfully (Davidon 1959; Fletcher and Powell 1963). The solution of the above system of equations is obtained with transformation in which the constrained equations are transformed into a sequence of unconstrained problems using the penalty concept. The essence of this method to form a barrier along the boundary of the feasible region. Several constrains simply limit the range of variables (cutting speed, feed sped, tooth bite etc.). The limit for waviness, roughness or edge radius will also be prescribed in the form of constraints.

It must be stressed that the appropriate selection and accomplishment of constraints requires special care. The lack of an important constraint may cause erroneous and unacceptable results.

The above strict mathematical method may be quite complicated even for simple problems. The following worked example will show the essence of this method. Let us consider a simple clamped cantilever made of wood and loaded on the free end with a force F (Csanády et al. 2019). This problem in the mechanics is well-established but we ask now the following: what is the shape of cross-section of a beam which has minimum volume for a given material, length and load provided that the cross-section is constant along the beam (Fig. 9.1).

The objective function, with the variables h and b, simply reads

$$V \approx F(x_1, x_2) \approx L \cdot b \cdot h = L \cdot x_1 \cdot x_2 = \min$$

Fig. 9.1 Cross-section of the beam

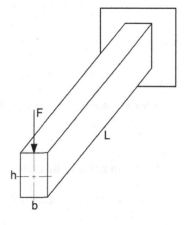

where $b = x_1$ and $h = x_2$.

The feasible region of the variables is delimited by the following constrains:
The first constraint is the allowable bending stress on the beam

$$h_1(x) = \sigma_b - \sigma(x) \geq 0 \text{ and } \sigma(x) = \frac{6F \cdot L}{b \cdot h^2} \tag{9.1}$$

with $\sigma_b = 2000 \text{ N/cm}^2$.

If the ratio $h/b = x_2/x_1$ becomes large, it involves the risk of the beam to buckling. Therefore, those combination of x_2 and x_1 that will cause buckling must be disallowed. The critical buckling load can be calculated as:

$$F_{cr} = \frac{4}{L^2} \frac{b^3 h}{6} \sqrt{E \cdot G} \left[1 - \frac{h}{4L} \sqrt{\frac{E}{G}} \right] \tag{9.2}$$

where the modulus of elasticity E and the shearing modulus G must be taken in the h-b plane ($E = 150{,}000 \text{ N/cm}^2$, $G = 40{,}000 \text{ N/cm}^2$). Using the given material property values, the critical load may be well approximated by the equation

$$F_{cr} \cong 49{,}319 \frac{b^3 h}{L^2} \tag{9.2a}$$

and the constraint is given by

$$h_2(x) = F_{cr}(x) - F \geq 0$$

Any of these constraints is omitted, the following impossible result would be obtained: $h = \infty$, $b = 0$, $V = 0$.

The allowable deflection of the beam may be also constrained in the following form:

$$f = \frac{4F \cdot L^3}{Ebh^3} \text{ with } E = 1.5 \times 10^6 \text{ N/cm}^2 \tag{9.3}$$

and

$$h_3(x) = f_{max} - f(x) \geq 0 \quad (f_{max} = 0.45 \text{ cm})$$

A nonnegativity restriction on h gives the last constraint

$$h_4(x) = x_2 \geq 0$$

The formulation of function to the problem in an unconstrained form yields

$$F(x) = 50x_1x_2 + R\langle h_1(x)\rangle^2 + R\langle h_2(x)\rangle^2 + R\langle h_3(x)\rangle^2 + R\langle h_4(x)\rangle^2 \tag{9.4}$$

where R is a set of penalty parameters and $\langle \rangle$ is the bracket operator with the following meaning

$$\langle a \rangle = a \text{ if } a \leq 0$$
$$\langle a \rangle = 0 \text{ if } a > 0$$

Using the Fletcher-Powell algorithm, the following solution will be obtained using increasing penalty parameter R. $b = 1.317$ cm, $h = 5.009$ cm and the minimum volume is 330 cm^3.

The numerical method supply only one specific solution and, changing the length or load, the numerical calculation should be repeated.

In the following we show an analytical approach, supplying the same results with the advantage that it can simply be extended to arbitrary conditions. It is important to recognize that the main constraint is the critical buckling load. Therefore the following basic equations will be used.

From Eq. (9.1) with $\sigma_b = 2000$ N/cm^2 we get

$$F = 333.3 \cdot \frac{bh^2}{L} \tag{9.1a}$$

If we take the condition $F_{cr} = F$, then, using Eq. (9.2a), the optimum beam height has the value

$$h = 148 \cdot \frac{b^2}{L} \text{ or } b = \frac{\sqrt{h \cdot L}}{12.16} \tag{9.5}$$

Multiplying Eqs. (9.1a) and (9.2a) yields

$$F_{cr} \cdot F = 1.6433 \times 10^7 \frac{b^4 h^3}{L^3}$$

Extracting a root from the above equation gives

$$F = 4054 \frac{b^2 h^{3/2}}{L^{3/2}} \text{ or } h = \frac{1}{254}\left(\frac{F \cdot L^{1.5}}{b^2}\right)^{2/3} \tag{9.6}$$

From Eqs. (9.5) and (9.6) the optimum value of b and h can uniquely be determined. Combining of these two equations, we obtain an equation for h and b:

$$h = 0.266\left(F \cdot L^{0.5}\right)^{0.4}$$

and

$$b = 0.0157 \frac{\left(F \cdot L^{1.5}\right)^{0,5}}{h^{0.75}} \qquad (9.7)$$

From these equations we can direct calculate the optimum values: $h = 5.03$ cm, $b = 1.305$ cm and $V = 328$ cm^3. These values agree with the numerical solution within the error of numerical calculations.

This worked example proves that there are more simple engineering methods which may give generally valid analytical solution. It is true, however, that the main constraint should be recognized.

It is worth to mention that the optimal column problem of Lagrange, which has been formerly solved with numerical method (Carter and Ragsdell 1974), has generally valid analytical solution (Csanády et al. 2019).

9.4 Application of Engineering Optimization Method

The different wood machining processes can be regarded as a particular subsystem of the whole manufacturing process which can be handled without greater difficulties. It may generally be supposed that an optimal subsystem is seldom in conflict with the whole production system. Nevertheless, in the case of conflict (i.e. synchronization of a production line) the required correction can be accomplished with good compromise.

In the following we demonstrate the application of engineering optimization method for selected problems of wood machining.

9.4.1 Primary Wood Processing

The main task of the primary wood industry to cut the raw material in to lumber with a given cross-section depending on the log diameter and the need of consumers. The requirement for surface quality is not very high but the cutting accuracy is important to reduce necessary allowance for subsequent woodworking operations. Surface irregularities depend on the blade stability, uneven tooth setting, low lateral stiffness and high feed speed (see Chaps. 5 and 6). The sawing performance directly depends on the feed speed and, therefore, the blade stability is of great importance.

The performance and energy consumption are determined by the cut cross-section in the unit time, which is calculated as the feed speed multiplied by the height of the sawn cross-section $e.H$. The maximum allowable feed speed is constrained by the gullet feed index which is the ratio of the loose chip volume and gullet feed volume (gullet area x kerf width). The chip produced has to fit in the gullet until its exit from the wood. The maximum allowable feed speed is theoretically equal to

$$e_{max} = \frac{V_g \cdot v}{\varepsilon \cdot V_w} \text{ with } V_w = H \cdot t \cdot b \tag{9.8}$$

where

V_g, V_w are the gullet and solid wood volume,
ε is the volume ratio for chip and solid wood which is generally around 3,
t is the tooth pitch,
b is the kerf width.

There is hyperbolic relation between feed speed and cross-section height: as the sawn height increases the allowable feed speed decreases. The loose chip can be slightly compacted with a small pressure. Therefore, the theoretical allowable feed sped may be increased by 20% without the risk of developing higher lateral pressure in the gullet causing additional friction. Figure 9.2 shows the relationship between feed speed and sawn height for a band saw cutting oak logs with different gullet feed indexes and also measurement results in a sawmill. An experienced worker could accurately select the appropriate feed speed according to the diameter of the sawn log (Déry 1985). The optimum performance of a sawmill mainly depends on the feed speed and minimization of non-productive time. Due to the limited feed speed, the optimum cost of the mill is obtained at the maximum allowable feed speed and maximum share of productive time. The latter requires skill and good organization of individual operations. Smaller sawn boards have a greater surface compared to their volume. Cutting small logs into boards or small pieces considerably increases production costs. The specific cutting surface of logs from 30 to 45 cm in diameter is approximately 23 or 37 m^2/m^3 if sawing $1''$ or $2''$ boards.

Experiments were conducted on a smaller bandsaw with wheel diameter of 1100 mm, a gullet area of 218 mm^2, a 30 mm tooth pitch and with a cutting speed of 30 m/s. The theoretical feedspeed was e = 4.36/H where the cutting height H must

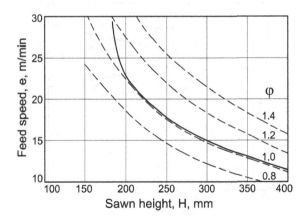

Fig. 9.2 The relationship between feed speed and sawn height for different gullet feed indexes φ. Solid line shows measurement results in a sawmill for oak, $v = 30$ m/s, $t = 30$ mm, $V_g = 500$ mm^3, $\varepsilon = 3$

be substituted in m. The bandsaw cuts in one direction and the backward motion occurred at a speed of around 40 m/min. The coefficient of time use was around 80%. Oak logs were sawn into 1″ (26 mm) boards using logs from 20 to 45 cm in diameter and of 4 m long (Déry 1985).

Figure 9.3 shows the specific energy consumption, without edge cutting, as a function of log diameter.

The board/log volume ratio fluctuates between 70 and 75%. The sawn volume moderately increases with the log diameter. Accordingly, the specific cost decreases to the same extent, Fig. 9.4. The following cost components are taken into account: machine cost $80/h, cost of three workers is $30/h and the tool cost with change is $15/h (Csanády et al. 2019).

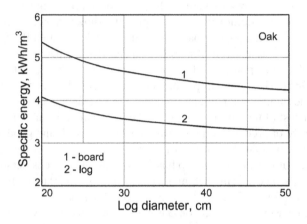

Fig. 9.3 Specific energy consumption of a bandsaw as a function of log diameter cutting 1″ boards

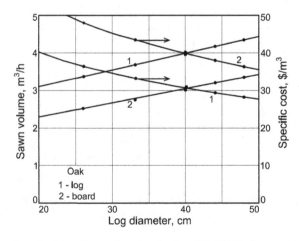

Fig. 9.4 Sawn volume and specific cost for a band saw cutting 1″ boards

Figure 9.4 clearly shows that, doubling the log diameter, the sawn volume in the unit time increases only some 30% and the specific cost decreases to the same extent. This is due to the fact that in each pass only one board is cut and, at the same time, with an increasing diameter the feed speed decreases. With increasing diameter, the number of boards and passes also increases and only the smaller cutting heights toward the edge of the log can be utilized. The quick return motion is an important time saving operation. Therefore some big bandsaws cut forward and back using a blade with teeth on the each side. Another possibility is to use higher cutting speeds as far as technically possible.

Frame saws have been used for a long time in European sawmills. In spite of its mature construction, the frame saw has also serious limitations, e.g. its rotation speed is limited to around 340 rpm. As a consequence, the feed speed is also limited and it may vary only as a function of cutting height. This drawback is, however compensated by using multiple cutting blades. The typical sawing pattern is shown in Fig. 4.7 in Chap. 4.

The total height of the sawn cross-section is calculated according to Eq. (4.4) depending on the log radius and board thickness.

The total sawn height approximately depends on the log diameter with an exponent of 2.2. For example, sawing board thicknesses of a = 7.5 cm and c = 4 cm wide, the total sawn height is

$$H_\Sigma = 0.07 d^{2.2} \text{ cm}$$

while for a = 5.2 cm, and c = 2.6 cm

$$H_\Sigma = 0.112 d^{2.2} \text{ cm}$$

where the log diameter d must be substituted in cm.

The theoretical feed speed (under the assumption that the chip has to fit in the gullet) can be calculated as

$$e = \frac{A_g \cdot v_a \cdot 60}{2 \cdot \varepsilon \cdot H \cdot t} \text{ m/min}$$

with the average cutting speed

$$v_a = \frac{D \cdot n}{30} \text{ m/s}$$

where

A_g is the gullet area,
t is the tooth pitch,
D is the stroke of the frame saw,
H is the maximum sawn height (log diameter).

Using the common values with $v_a = 6.5-6.8$ m/s, $t = 25-30$ mm, $A_g = 2.4-2.8$ cm^2, the theoretical feed speed is

$$e = \frac{0.63\text{--}0.76}{H} \text{ m/min} \tag{9.9}$$

where the sawn height H must be substituted in m. In practice this feed speed may be increased with about 20% to the cost of chip compaction in the gullet space without the perceptible increase of friction forces.

The pure cutting power is proportional to the sawn cross-section in the unit time, eH_Σ in m^2/min. This measurement is depicted in Fig. 4.6 for Scotch pine and black locust. Similar results of more general validity are given in Fig. 4.15 (Chap. 4).

Keeping in mind that Eq. (9.9) may increase, the volume of sawn logs in the unit time is proportional with the log diameter and the efficiency of time utilization η_T

$$V \cong 45 \cdot d \cdot \eta_T \text{ m}^3/\text{h} \tag{9.10}$$

where d must be substituted in m.

The specific energy consumption (kWh/m^3 log) depends on the sawing pattern, wood species, log diameter and the power consumption of the frame saw when it is idling.

The idling power consumption of frame saws with a hydraulic drive is relatively high, around 20 kW. Figure 9.5 shows measurement results in a sawmill cutting Scotch pine and using two sawing patterns (Sitkei et al. 1988). The time utilization was 80% and the energy consumption is only from the main cutting without edge cutting. The latter was made by a double circular saw with an adjustable cutting distance.

The specific energy consumption was calculated for Scotch pine as

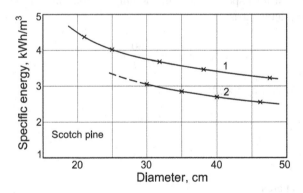

Fig. 9.5 Specific energy consumption of a frame saw cutting Scotch pine, using two sawing patterns as a function of log diameter. 1. $a = 5.2$ cm, c = 2.6 cm 2. $a = 7.5$ cm, c = 4 cm

$$\frac{P}{V} = \frac{(1 - \eta_T) \cdot P_0 + \eta_r \cdot (P_0 + 4.1 \cdot e \cdot H_\Sigma)}{\eta_T \cdot 45 \cdot d} \quad \text{kWh/m}^3 \qquad (9.11)$$

with $P_o = 20$ kW idling power and $\eta_T = 0.8$.

The production of boards from the raw material may be as high as 70% if the logs are properly sorted according to their diameter. Accordingly the specific energy consumption related to the volume of boards will be 43% higher compared to Fig. 9.5.

Using double circular saws for edge cutting, each saw requires a maximum driving power of 10 kW in order to use feed speeds up to 70–75 m/min for boards one or two inches thick. Cutting 3-inch boards, the feed speed should be reduced to appr. 50 m/min. The specific energy requirement for edge cutting fluctuates between 0.45 and 0.3 kWh/m^3 board as a function of log diameter (conifers). Hardwoods (oak, black locust, beech) require 40–50% more energy for sawing (see Fig. 4.6). The energy requirement and its components as a function of log diameter is demonstrated in Fig. 9.6 which is related here to the unit board volume. 70% of the log volume is turned into boards (Csanády et al. 2019).

The specific sawn surface area and its components vary as a function of log diameter, using a given sawing pattern, Fig. 9.7 (Csanády et al. 2019). The sawn surface area of boards cut slightly increases while that of the edge cut slightly decreases with increasing log diameters. As a result, the total sawn surface does not vary much and its average value is roughly 45 m^2/m^3 board for each sawing pattern.

Fig. 9.6 Specific sawing energy related to the board volume and its components as a function of log diameter (Conifers). Sawing pattern: $a = 52$ mm, $c = 26$ mm

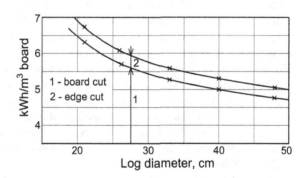

Fig. 9.7 Specific sawn surface area related to the board volume and its components as a function of log diameter. Sawing pattern: $a = 52$ mm, $c = 26$ m

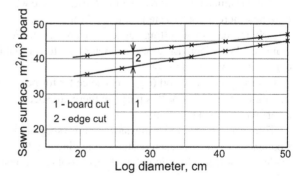

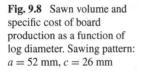

Fig. 9.8 Sawn volume and specific cost of board production as a function of log diameter. Sawing pattern: $a = 52$ mm, $c = 26$ mm

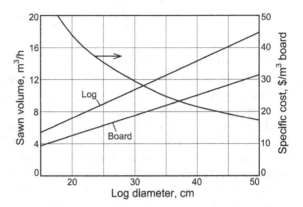

Due to the big difference in the feed speeds for frame saws (2–4.5 m/min) and circular saws (50–70 m/min), both sawing operations can be synchronized in time and the edge cut does not influence the capacity of the frame saw. The organization of material handling is very important because the volume of sawn logs in the unit time directly depends on the efficiency of time use, as seen in Eq. (9.10).

The specific cost of the boards ($/m^3) depends on the various cost elements such as the machine, labour, energy and overhead costs, tool and sharpening costs, and further the setting, tool changing and nonproductive times. Taking the prime cost of all the machines as $900,000, the machine costs $160/h, the labour for 3 workers costs $30/h and the tool and tool changing cost 30 $/h, the total cost in the unit time is $220/h. Using the above example given in Figs. 9.6 and 9.7, the sawn volume and its specific cost as a function of log diameter are shown in Fig. 9.8 (Csanády et.al. 2019). Clearly the log diameter influences the unit cost considerably.

The best way to reduce costs is good organization of subsequent operations, decreasing the non-productive time. It is to ensure the highest possible board outcome using log sorting and an appropriate sawing pattern.

The energy cost of sawing is not very significant compared to other costs. The cost of 1 kWh energy is appr. $0.10 and therefore the net energy cost is around $0.50–$0.60/m^3 board (see Fig. 9.6).

Due to the strong limitation of system variables (cutting and feed speed), optimization is much more a striving after the ideal case where there is no non-productive time and there are no waste products. At the same time, the use of multiple blades linearly increases the cutting capacity, according to Eq. (9.10), as a function of log diameter. With increasing log diameters up to its maximum allowable it is fully competitive with a bandsaw.

Cutting precious wood (e.g. incense cedar for pencils) or thin lamellae, thin kerf sawing is necessary. For this purpose there are special frame and circular saws with a kerf width down to 1.2 mm.

Thin wall circular and frame saws require special preparation (tensioning, natural frequency), often after each resharpening. Sawing deviations strongly depend on the blade thickness as a function of feed speed as it is demonstrated in Fig. 5.22 (Chap. 5).

9.4.2 General Wood Machining

The sizing, shaping and surface preparation of wood products occur also in our days with mechanical woodworking operations. The mechanical woodworking is characterized by the chip formation accomplished by a cutting edge (see Sect. 1.3). The working principle of the cutting edge is very similar in the different woodworking operations but the arrangement of edges on tools shows a great diversity. An exception is the abrasive sanding process using negative rake angle and random position of its grits. Wood cutting tools utilize the free cutting mechanism without counteredge. The required counter force is ensured by the strength of the material and inertia forces. In order to achieve a clear cut without surface deformations, high cutting speed and sharp edge are required. Soft woods have smaller strength and they are more sensitive to lower cutting velocities (see Figs. 8.20 and 8.21). Veneer knifing machines have lower cutting velocity and in order to get an acceptable surface quality, the use of pressure beam is unavoidable.

A very important task is the proper selection of tools, their edge material and configuration, which are in close relation with the expected tool life and surface quality.

A wide range of tungsten carbide tool materials are available to the woodworking industry for many different uses. The different carbide grades are composed of various percentages of tungsten carbide and a metallic binder (cobalt). Furthermore, these tool materials utilize a range of carbide grain sizes. The most common grain size is about 2 microns but there are grades in the sub-micron range (Feld et al. 2005; Garcia 2005). Varying the carbide-binder ratio and the grain size, tool materials with different mechanical properties are made to ensure optimum performance when machining different types of wood and composites (hard and softwoods, MDF, particleboard etc.).

The economy of manufacturing considerably influenced by the woodworking operations including knife machining and sanding. Different criteria (objective functions) may be posed to achieve maximum production rate, minimum production costs, but in some cases the maximum tool life may also be important. An array of constraints must always be taken into account. Constraints of the machining parameters are

- the available feed rates,
- the available rotation speeds and cutting speeds,
- the maximum power or torque available,
- stable cutting area excluding certain combinations of cutting speed and feed speed causing vibrations and tool instability.

Technological requirements are another class of constraints which may be regarded as *goal constraints*:

- the allowable waviness of the surface,
- the allowable maximum surface roughness specified by a single or several roughness parameters,

- required size accuracy,
- the material removal rate (MRR) should be specified or constrained to a minimum rate,
- the tool life may be specified to achieve optimum time between tool changes or constrained to a minimum tool lifetime.

The Maximum Production Rate requires a maximum feed rate. The feed rate is in a simple relation with the main operational parameters as

$$e = e_z \cdot n \cdot z \text{ m/min}$$

where

e_z is the tooth bite,
n is the rotation speed,
z is the number of teeth.

We may use the waviness t_1 as a prescribed constraint which is given by

$$e_z = 2\sqrt{D \cdot t_1}$$

where D means the tool diameter, then the feed speed is calculated as

$$e = 2 \cdot \sqrt{D \cdot t_1}(n \cdot z) \tag{9.12}$$

which is plotted in Fig. 9.9 for different tool diameters and at a waviness of 2 μm (Csanády et al. 2019).

The production rate depends on the material removal rate (MRR), which is the material volume removed in the unit of time

$$V = e \cdot H \cdot b = 2\sqrt{D \cdot t_1} \cdot H \cdot b \cdot (n \cdot z) \text{ cm}^3/\text{min} \tag{9.13}$$

where H is the depth of cut, and b is the width of cut.

The width of cut may often vary and, in this case, it is more convenient to relate the material volume to the unit width

$$\frac{V}{b} = 2 \cdot \sqrt{D \cdot t_1} \cdot H \cdot (n \cdot z) \text{ cm}^3/\text{min cm} \tag{9.13a}$$

Taking $D = 10$ cm and $t_1 = 2$ μm, Eq. (9.13a) is plotted in Fig. 9.10 which allows a quick estimate for different conditions.

The next important constraint may be a prescribed upper limit for roughness parameters. As explained in Sect. 8.5, the average roughness parameter R_z depends on the feed per tooth e_z in the following fashion (Fig. 8.24):

Fig. 9.9 Feed speed as a
function of rotation speed
x number of edges for
different tool diameters

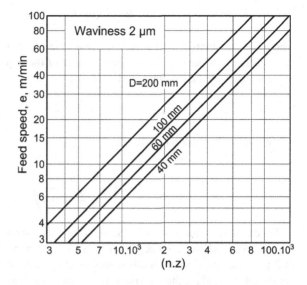

Fig. 9.10 Removal rate as a
function of rotation speed
x number of tool edges for
different depth of cut. $D =$
100 mm, $e_{zmax} = 0.9$ mm

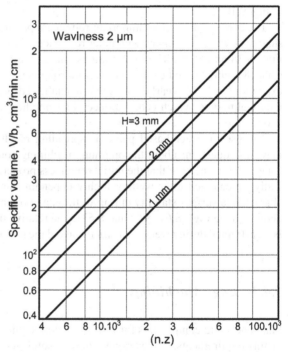

$$R_z = A + B \cdot e_z^n \ \mu m$$

where the constants B and n may generally be taken as $B = 43$ and $n = 0.6$, but the constant A is determined by the anatomy of the given wood species. It may be express either by the first term of Eq. (8.16)

$$A = 123 \cdot \Delta F^{0.75}$$

or with the Abbott-ratio (see in Fig. 8.22)

$$A = \left(\frac{38}{\frac{R_{pk}+R_k}{R_{vk}} + 0.4} \right)^{1.05}$$

For example, when machining Scotch pine we limit the roughness to $R_z = 40 \ \mu m$, the feed per tooth must not be higher than $e_z = 0.5$ mm. The constraint for waviness of 2 μm allowed feed per tooth values up to 0.9 mm and, therefore, the roughness constraint will reduce the material removal rate in a ratio of $0.5/0.9 = 0.55$ compared to Fig. 9.10.

The core depth R_k is an important roughness parameter and an indicator for tool sharpness which correlates with other roughness parameters such as R_a and R_z (see Figs. 8.38 and 8.39). The interrelations among the roughness parameters given in Chap. 8 highly facilitate the use of various roughness parameters as a constraint. For example, Fig. 8.24 clearly shows that a small feed per tooth considerably reduces the roughness R_z which is the essence of a smoothing pass and has been known for a long time.

The use of water based paints is spreading results in a **wetting process** and may raise questions about the optimum machining conditions for this purpose (see Chap. 8). It turned out that the initial roughness properties of a surface may dramatically be changed after wetting highly depending on the kind of machining. To characterize surface stability during wetting, a good indicator is the use of the core depth R_k (see Figs. 8.61, 8.62 and 8.63). The relative increase of this indicator may be constrained during wetting in order to avoid the necessity of intermediate sanding.

9.4.3 Edge Machining

Flat panels and components for furniture often require narrow face machining with quality requirements. The workpiece may be solid wood or particle board, laminated or not. The narrow surface is generally 10–20 mm wide and the edges are very sensitive to breaking which considerably lowers the quality and aesthetical value of the workpiece. The edge break phenomenon may be explained by the fact that the cutting force originates spatial stress distribution in the board but the two flat surfaces are not supported. Especially tension stress in the edge promote edge breaking.

Fig. 9.11 Moulder edge machining with orthogonal cutting (**a**), two reversible knives with overlap and oblique angle $\lambda = 8°$ (**b**), two moulders with setting angle, counter cutting

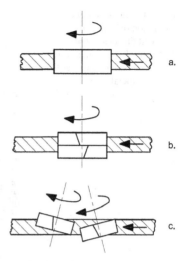

Number and size of edge breaks determine the quality of narrow face machining, (see in Fig. 8.4a). The characteristic measures are the length (l_z), the depth (l_y) and the width (l_x) of breaks (notches). A special perthometer is developed in which, instead of needle, a wedge traces the edge (Saljé and Drückhammer 1985). A quality number is defined which sums up the weighted elementary cross-sections of the breaks as a function of depth (see in Fig. 8.5).

For narrow face machining different moulding systems can be used. The most common tool is the peripherical moulder with orthogonal cut Fig. 9.11. A better performance is achieved using two reversible knives with overlap and oblique angle (Fig. 9.11b). A more expensive solution utilizes two cutterheads: with setting angle (Fig. 9.11c). The last two solutions have the advantage that the forces on both side are inwards directed and excert compressive stress onto the edge.

In general, edge breaking can be reduced by using a larger tool diameter and thin chips. The use of cone-shape cutters fulfils both requirements (see Sect. 3.5).

Using the common milling cutters, edge breaking is influenced by the tooth feed e_z and mainly by the sharpness of the edge or feed distance L_f. Figure 9.12 (Licher 1991).

Using the results given in Fig. 9.12, the maximum feed distance L_f can easily be calculated for different conditions. For example, if the allowable shardness is 1 mm^2/m then the maximum feed distance per edge is $L_f = 82.1$ m at a tooth bite of $e_z = 2.31$ mm which corresponds to 677 m cutting distance per edge and a feed speed of $e = 14.7$ m/min. If the allowable shardness is doubled ($S = 2$ mm^2/m), then the corresponding values are: $L_f = 179.8$ m, $L_c = 1007.4$ m, $e_z = 3.4$ mm and $e = 21.6$ m/min per edge.

A further analysis of the obtained results revealed that the optimum values of L_f and e_z to the selected shardness S have the following relationship

Fig. 9.12 Edge shardness as
a function of feed per tooth
and feed distance. $D =$
180 mm, $H = 2$ mm, $z = 1$,
$v_c = 60$ m/s. Particle board
and hard metal edge

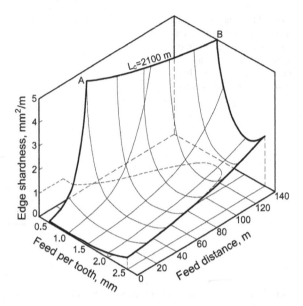

Fig. 9.12 Edge shardness as a function of feed per tooth and feed distance. $D =$ 180 mm, $H = 2$ mm, $z = 1$, $v_c = 60$ m/s. Particle board and hard metal edge

$$\left(\frac{L_f}{e_z}\right)_{opt} = \frac{const \cdot S^{0.59}}{\sqrt{R \cdot H}} \tag{9.14}$$

Since the optimum tooth bite only depends on the selected shardness S, that is

$$(e_z)_{opt} = const \cdot S^{0.55} \tag{9.15}$$

the optimum feed distance is calculated as

$$\left(L_f\right)_{opt} = \frac{const \cdot S^{1.14}}{\sqrt{R \cdot H}} \tag{9.16}$$

Using Fig. 9.12 with a cutting speed of 60 m/s, constant in the above equation is 1.1 if R and H is substituted in m and the shardness in mm^2/m.

This last equation also indicates that the shardness is primarily determined by the edge wear. It is interesting to note that tool radius and cutting depth considerably influence the optimum feed distance but not the optimum tooth bite which depends only on the allowable shardness.

Using tools with multiple edges, the feed distance increases according to the number of edges. In the above example, due to the high cutting speed of 60 m/s, the pure cutting time between two sharpenings is relatively short. Calculating for different allowable shardness values, the cutting time T may be expressed by a simple power equation ($R = 90$ mm, $H = 2$ mm):

$$T = \frac{L_f}{e} = \frac{82.1 \cdot S^{1.13}}{14.7 \cdot S^{0.55}} = 5.58 \cdot S^{0.58} \text{ min}$$

where the shardness S must be substituted in mm^2/m.

Similar experimental results to Fig. 9.12 are not known for other cutting speeds. Due to the governing role of edge wear, the tool-life equation allows an approximate analysis of effect of cutting speed. Figure 7.7 shows a similar case (particle board and hard metal edge) with the tool-life equation

$$T \cdot v^{2.2} = const.$$

or

$$T_1 v_1^{2.2} = T_2 v_2^{2.2}$$

If we lower the cutting speed from 60 to 40 m/s, then the cutting time increases as follows:

$$T_2 = \left(\frac{v_1}{v_2}\right)^{2.2} \cdot T_1 = 2.44 T_1$$

$$e_2 = \frac{v_2}{v_1} \cdot e_1 = \frac{2}{3} \cdot e_1$$

$$L_{f2} = e_2 T_2 = 1.627 \cdot e_1 T_1$$

where e_1 and T_1 are the known values for feed speed and cutting time at 60 m/s cutting speed. The cutting time T is more than doubled, the feed speed decreases by one third and the feed distance increases some 60%. This may profitably aid the organisation of tool change and tool costs.

The specific material removal for a given cutting speed v and feed rate e is calculated as follows

$$V_T = \frac{L_{fT} \cdot H \cdot b}{T} \text{ m}^3/\text{min} \tag{9.17}$$

Changing the cutting velocity to v_x, the removal rate varies according to the following equation

$$V_{Tx} = \frac{e \cdot H \cdot b}{(v/v_x)} \text{ m}^3/\text{min}$$

In the previous example with $v = 60$ m/s, $v_x = 40$ m/s and $e = 14.7$ m/min, the material removal rate decreases according to the velocity ratio of 2/3. That means that the production rate is in conflict with decreasing cutting speeds. The frequent tool change and sharpening cost, however, may influence the production cost in an other manner.

The volume of material removal (m^3) between two sharpening is, however, increasing with decreasing cutting velocity as follows

Fig. 9.13 Interrelation
between optimum tooth bite
and distance for different
shardness values and the
effect of tool diameter and
depth of cut

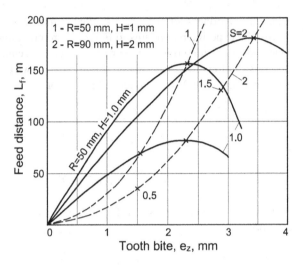

Fig. 9.13 Interrelation between optimum tooth bite and distance for different shardness values and the effect of tool diameter and depth of cut

$$V_{Tx} = V_T \cdot \left(\frac{v}{v_x} \right)^{1.2} m^3$$

but the time T between two sharpening increases in a higher extent

$$T_x = T \cdot \left(\frac{v}{v_x} \right)^{2.2}$$

which explains the decrease of specific material removal with decreasing cutting speed.

Using Eqs. 9.15 and 9.16, a chart can be elaborated such as shown in Fig. 9.13 which can easily be used to make some compromise in the selection of tooth bite, tool diameter and depth of cut. The feed distance curve in the vicinity of optimum tooth bite may be regarded as flat which allows some deviation from the optimum value with minimum loss in feed distance. The tool diameter and depth of cut have considerable effect on feed distance but do not influence the optimum tooth bite, as mentioned already before.

9.4.4   Machining of Curved Surface

High quality furniture, decorative relief surfaces and wooden artefacts require the production of curved surfaces. Curved surfaces may be two or three-dimensional depending on the complexity of the design pattern. A typical example is the curved door of a cabinet which is curved horizontally but straight vertically (two-dimensional surface). Furthermore, the curved surface locally may be convex or concave, or both.

Curved surfaces can be manufactured in different ways. Bending and pressing technologies are often used for bending and pressing plywood sheets to a given profile (sitting surfaces for seats). These technologies are, however, limited to produce moderate curvatures and simpler profiles.

Curved surfaces on solid wood base can be manufactured by machining. Axisymmetric pieces may be made on turning lathes which are controlled by a template or original workpiece. Profiled turning knives or sawblade like cutter are used according to the required profile. They were formerly served manually but today there are semi-automatic and fully automatic turning machines.

The produce curved surfaces of arbitrary profile, CNC machining centres are especially suitable due to their ability to change tools according to the requirement to achieve maximum efficiency in the production rate and cost. Generally a 5-axis CNC machine is required if different tool-surface inclination is needed.

The main problem in the production of curved surfaces is the inherent geometric roughness which should be kept as small as possible. Geometric roughness is the result of interaction of the tool profile with the surface. Generally a ball-end milling is used and the main process variables are the tool profile selection, the tool path, tool-surface inclination and the cutting conditions (cutting speed, feed rate, tooth bite, depth of cut, and number of passes). A multi-tool milling operation is used when more than one operation will be done, such as face milling, corner milling, pocket milling or slot milling. Multi-pass milling first does rough machining followed by finish machining.

To machine curved surfaces, it is important to select the correct tool profile. Figure 9.14 shows an example of machining a half-cylinder surface with ball-end milling. If the radius of the workpiece is much greater than the radius of tool, then a plain substitute model may be used with simpler relations.

The height of geometric roughness z can be obtained from the following equation:

$$r^2 = \left(\frac{b}{2}\right)^2 + (r - z)^2$$

where r is the tool radius and b is the distance between two adjoining passes. The above equation yields

$$z = r - \sqrt{r^2 - \frac{b^2}{4}} \qquad (9.18)$$

or in dimensionless form

$$\frac{z}{b} = \frac{r}{b} - \sqrt{\left(\frac{r}{b}\right)^2 - \frac{1}{4}}$$

This equation clearly shows that the cutting width b and the tool radius r fundamentally influence the geometric roughness z. The plain substitute model does not give a correct result if the tool radius tends to infinity; i.e. if the tool has a flat face.

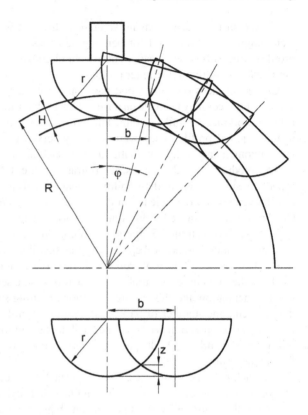

Fig. 9.14 Milling of a half-cylinder surface and its substitute model

In the latter case, the geometric roughness is determined by the local radius R of the machined surface and by the cutting width b:

$$z_0 = R \cdot \frac{1 - \cos\varphi/2}{\cos\varphi/2} \quad \text{and} \quad \varphi = \frac{b}{R} \text{ radian} \qquad (9.18a)$$

which is equivalent to

$$z_0 = 0.125 \cdot \frac{b^2}{R} \qquad (9.18b)$$

This equation can be accurately converted into the simpler equation

$$\frac{z}{b} = \frac{0.125}{\left(\frac{r}{b}\right)}$$

or

$$z = 0.125 \cdot \frac{b^2}{r} \qquad (9.18c)$$

It is obvious that the geometric roughness cannot exceed the depth of cut H, therefore

$$z \leq H \text{ or } z_0 \leq H$$

and that is always fulfilled if the maximum width of cut is limited to

$$b_{max} \leq \sqrt{\frac{H \cdot r}{0.125}} \text{ or } b_{max} \leq \sqrt{\frac{H \cdot R}{0.125}}$$

for ball-end milling and flat face milling respectively. Convex, axi-symmetric tool profiles, such as conical or spherical, have the inherent drawback that their cutting velocity in the centre line is zero. As a consequence, they need a higher feed force, produce more surface roughness in the vicinity of the centre line with low cutting speeds, especially when the feed direction is perpendicular to the grain. Soft woods are more sensitive to low cutting speeds than hard woods and using a sharp cutting edge and lower rake angles (around 15°) may help to overcome the problem.

A better solution is the selection of a proper tool-surface inclination angle, Fig. 9.15. In this case the centre line and its vicinity does not take part in the cutting of final surface. If the allowable geometric roughness is prescribed, then the width of cut b may not be higher than

$$b_{max} = \sqrt{\frac{z \cdot r}{0.125}}$$

For example, if $z = 0.1$ mm and $r = 30$ mm, then $b_{max} = 4.9$ mm.

To avoid the cutting action of the axisymmetric point, the minimum inclination angle should be

Fig. 9.15 Using inclination angle in ball-end milling

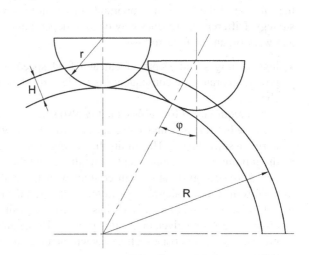

$$\varphi_{min} = \arccos\left(1 - \frac{z}{r}\right)$$

or its accurate substitute function

$$\varphi_{min} = 81.7 \cdot \sqrt{\frac{z}{r}}$$

In practice it is worth to select a somewhat higher inclination angle which will exclude the low speed part of the tool from the cutting process.

Using multi-pass machining, in the rough milling passes higher geometric roughness z is allowed and only the final smoothing pass is performed with the prescribed roughness.

The general rules of wood machining are also valid here, For instance, increasing feed speed increases roughness, especially when the feed direction is perpendicular to the grain. Climb cutting (down milling) is more demanding to edge sharpness than counter cutting (up milling).

It should be noted that on curved surfaces the grain direction varies from place to place and, as a consequence, the surface roughness may change in a higher extent. Sadly, the measurement of roughness on curved surfaces is more complicated and very few measurement results are available.

9.4.5   Lifetime of Tools

Due to the continuous wear process, the lifetime of a tool is finite and its change from time to time is needed. The lifetime of the tool is also an important cost factor including its prime cost and maintenance cost. Considering the objective function, the lifetime of the tool is an important constraint depending on the manufacturing strategy. Different approaches may be used depending on the machining conditions and work organisation as follows:

- high cutting speed, depth of cut and feed speed is selected for a maximum production rate. The consequence is a relatively short tool life with frequent tool change,
- a high requirement for surface quality also shortens the tool life,
- the edge lifetime is bound to one shift in order to allow workers to change the tool during a shift change. This limits the cutting speed to a certain extent,
- the minimum cost requirement generally limits the cutting speed and depth of cut. This strategy is always in conflict with the maximum production rate. A compromise may balance the difference between the two strategies.
- optimum selection of edge materials to ensure a desired production rate and cost effectiveness for a given machining process. Depending on the priority, either the production rate or cost effectiveness will be taken as an objective function while the other is taken as constraint.

The material removal rate (MMR) fundamentally determines the production rate which is often a key factor in optimization. MMR is given by the following simple relations

$$MMR = e \cdot H \cdot b \ \text{cm}^3/\text{min}$$

which is apparently independent of cutting speed. Naturally, in order to maintain the tooth bite in the feasible range, feed speed and rotation speed should properly be selected.

The cutting depth H decreases the tool life and the feed distance but the volume of material removal suppresses the decreasing effect of tool life reduction.

The material removal during the life time of the tool is given by

$$V_T = e \cdot H \cdot b \cdot T = L_{fT} \cdot H \cdot b$$

and, keeping in mind Eq. (7.3a), we get

$$V_T = \frac{e_z \cdot z \cdot b}{1.42 \cdot v^{n/m}} \cdot \sqrt{\frac{H'}{R}} \left(\frac{y - y_0}{A}\right)^{1/m} \text{m}^3 \tag{9.19}$$

which is an important relationship for maximization of production rate.

Using a small tooth bite or low feed speed, the tool life will be reduced in respect to the material removal. Using multipass machining, there may be economic to use several rough passes with a bigger tooth bite and the final pass with small tooth bite. A deep depth of cut and larger tool diameter also increases the relative value of a cutting length. The effect of depth of cut can be seen in Figs. 7.7 and 7.8 in Chap. 7.

The exponent of cutting speed is generally $n/m = 1.2$, so the material removal slightly decreases with higher cutting speeds due to the more intensive wear. The following example shows the use of the above equation.

Beech wood was used to conduct experiments as a function of feed distance (see Figs. 8.37, 8.38 and 8.39). The following tool and operational parameters were used, tool diameter $D = 120$ mm, number of teeth $z = 4$, rotation speed $n = 6000$ rpm, cutting speed $v = 37.7$ m/s, feed speed $e = 12$ m/min, tooth bite $e_z = 0.5$ mm and depth of cut $H = 2$ mm. The cutting speed was examined in the 3000–8000 rpm range. The hardmetal edge had a sharpening angle of $\beta = 50°$ and therefore the relation between the edge radius and theoretical edge reduction is $\rho = 0.73y$. Using Eq. (7.3), the edge wear after 1800 m feed distance has the following expression

$$\rho = \rho_0 + 0.0297 \cdot v_c^{0.65} \cdot L_c^{0.52} \tag{9.20}$$

or

$$y = y_0 + 0.034 \cdot v_c^{0.65} \cdot L_c^{0.52}$$

The initial values of ρ_0 and y_0 were 15 μm and 20.5 μm respectively while after 1800 m feed distance $\rho = 60$ μm and $y = 82.2$ μm. The total feed distance refers to four teeth and therefore for one individual tooth the feed distance is 450 m and the cutting distance is 14,000 m. Similarly to Eq. (7.6), the Taylor tool life equation has the form

$$v^{2.25} \cdot T = \left(\frac{\rho - \rho_0}{A} \right)^{1.92} \cdot \frac{0.0737}{\sqrt{H/R}} \qquad (9.21)$$

or substituting the corresponding values from Eq. (9.20), we get

$$v^{2.25} \cdot T = \frac{96.365}{\sqrt{H/R}}$$

or

$$v \cdot T^{0.44} = \frac{164}{(H/R)^{0.22}} \quad \rho = 60 \ \mu m$$

From Eq. (9.21) it is clear that the constant of the Taylor equation is determined by the wear limit which is here $\rho = 60$ μm. If a roughness constraint allows only a smaller tool edge radius, then the constants decrease and tool life will be shorter. Using a deeper cut, the cutting length increases and the tool life decreases. The roughness parameter R_z of beech wood in Fig. 8.39 is related to the edge radius with the empirical equation

$$R_z = 40 + 0.45 \cdot (\rho - \rho_0) \ \mu m \qquad (9.22)$$

where the edge radius ρ must be substituted in μm. Selecting a maximum allowable roughness value and using Eq. (9.21), the lifetime of the tool can easily be calculated.

Figure 9.16 shows the relationship between tool life and allowable edge radius for two cutting speeds which is near a quadratic function. Using different depth of cut, the tool life may be longer or shorter depending on the depth of cut compared to $H = 2$ mm used in Fig. 9.16. For example, in the case of $H = 1$ mm the tool life increased by $\sqrt{2} = 1.414$ compared to Fig. 9.16.

The cutting speed greatly shortens the tool life but a higher cutting speed also means a higher material removal rate. Therefore it is interesting to examine the true effect of cutting speed. The specific material removal between two sharpenings is expressed for our beech wood according to Eq. (9.19) as

$$\frac{V_T}{b} = \frac{e_z \cdot z}{1.42 \cdot v^{1.25}} \sqrt{\frac{H}{R}} \left(\frac{\rho - \rho_0}{A} \right)^{1.92} m^3/m$$

which is plotted in Fig. 9.17.

Fig. 9.16 Tool life versus allowable edge radius for different cutting speeds on beech wood

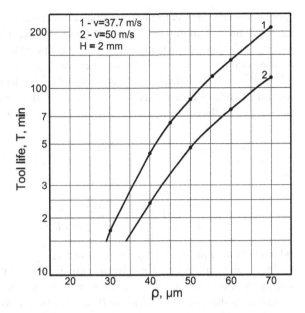

Fig. 9.17 Specific material removal rate as a function of allowable edge radius for beech wood

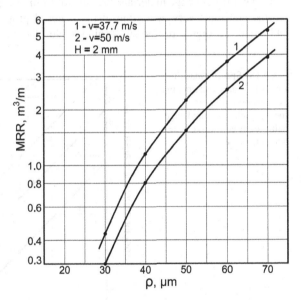

Using higher cutting speed, the material removal during the lifetime of the tool is decreasing with a factor of $(v_2/v_1)^{1.25} = 1.42$. Increasing the depth of cut, the material removal increases only with the square root of the H/R ratio.

Selecting the allowable surface roughness and the corresponding edge radius, they fundamentally influence the material removal during the lifetime of the tool. A tool edge radius that is too low substantially decreases the production rate. With a high roughness requirement it is worthwhile to make a final smoothing pass with a small depth of cut. For example, comparing a smoothing depth of cut $H_1 = 0.2$ mm to the common value $H_2 = 2$ mm, then the feed distance increases in a ratio of $\sqrt{H_2/H_1} = 3.16$ which is quite considerable. Due to the small depth of cut, the machined surface area related to the lifetime of the tool may be more useful. The area of the machined surface is proportional to the feed distance and the width of cut

$$A_{fT} = L_{fT} \cdot b$$

and

$$\frac{A_{fT}}{b} = L_{fT} = \frac{0.704 \cdot e_z \cdot z}{v^{1.25} \cdot \sqrt{H \cdot R}} \left(\frac{\rho - \rho_0}{A}\right)^{1.92} \text{ m}^2/\text{m}$$

which is plotted in Fig. 9.18 ($e_z = 0.5$ mm and $z = 4$).

Increasing cutting speed and depth of cut decreases the area that can be machined during the life of the tool. The allowable maximum edge radius is also a major factor determining the time a machine can be used between two edge sharpenings.

Using different edge materials, the general form of Eq. (9.20) remains unchanged but the constants vary according to the material properties of both the tool edge and wood. Furthermore, the sharpening angle of the tool has also some influence

Fig. 9.18 Specific machined area between two sharpening as a function of cutting speed, depth of cut and edge radius on beech wood, $e_z = 0.5$ mm, $z = 4$

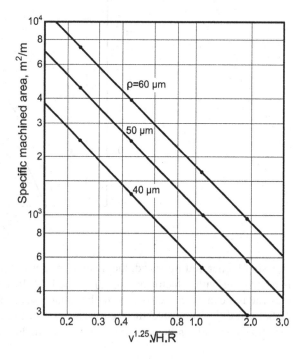

on the exponent of the cutting distance. A higher sharpening angle create less heat load therefore causes less wear resulting in a somewhat lower exponent. Composite materials containing adhesives increase the wear of a tool and shorten its life.

About the different hard metals, their properties and appropriate selections see Sect. 9.4.2. An optimum selection of edge material considerably contributes to a cost and time effective production.

9.4.6 Manufacturing Costs

Optimization of machining is an important part of an overall economic production but in an actual manufacturing process further additional time and cost elements are present.

The main time components are as follow:

- machine preparation (setup) time,
- loading unloading time,
- process adjusting and return time,
- machining time,
- tool changing time.

Using CNC machining centres, the setup time may be reduced dramatically with the concept of an external setup. This means that part of the setup operation which can be performed while the machine is operating on the previous series. Another way to reduce setup time is the clever design of tools and fittings enabling a quick change of tools and fittings.

Machining costs and other machine-related operation costs are considerable part of the total manufacturing cost. Due to the high capital investment, these machines need to operate as efficiently as possible. The costs of machining are following:

- machine cost per hour,
- labour cost per hour,
- tool cost, either for sharpening or changing knives, tool change cost,
- setup cost.

The machine and labour costs depend on the time required for machining.

The estimation of the loss of time for setup may be quite different. If the machine is not working at its full capacity, then a longer setup time will cost not too much. On the contrary, if the machine is a "bottleneck" constraining the total production, then the lost production due to setup time is a loss in production for the whole factory. Accordingly, the setup costs should be determined taking the real consequences into account. If necessary, one solution is the use of additional overtime work, which has its own extra cost.

The usual way of calculating the cost of the setup is to take the normal labour cost and the hourly cost of the machine without taking other cost elements into account. Therefore the simplified setup cost calculation is often inappropriate.

An important class of optimization is the selection of operational conditions to ensure the **minimum cost per component** manufactured. That is a cost optimization procedure and for machining, the cost and time elements between two sharpenings should be considered.

The time cycle may be defined as the sum of the following time elements

$$\sum t_i = t_s + T + t_{ch} \tag{9.23}$$

where

t_s is the setup time,
T is the tool life between two sharpening,
t_{ch} is the time for tool change.

During the tool life N component will be produced which is calculated as

$$N = \frac{T}{t_m + t_n} \tag{9.24}$$

where

t_m is the machining time for one component,
t_n is the loading and unloading time.

The calculation of machining time is based on the feed distance per component L_f^c and feed speed e in the following form

$$t_m = K \cdot \frac{L_f^c}{e} = K \cdot \frac{L_f^c}{e_z n \cdot z} = K \cdot \frac{2 \cdot \pi}{60} \frac{L_f^c R}{e_z v \cdot z} \tag{9.25}$$

where K is a correction factor taking the necessary approaching and return time into account ($K = 1.05$–1.1). The total cost related to a time cycle is given by

$$C = x[t_s + T + t_{ch}] + y \tag{9.26}$$

where

x is the labour and machine cost in the unit time,
y is the cost of sharpening or replacing the edge of the tool.

The machine cost in the unit time depends on the cost of amortisation, interest rate, maintenance cost, and space and energy costs. The tool cost includes the original cost of the tool and sharpening cost. The cost of a tool depends on its total lifetime while the sharpening cost is determined by the number of workpieces machined between two sharpenings. The number of sharpenings is given by the total lifetime of the tool and the tool life between two sharpenings.

An approximate machine cost per hour can ben estimated by taking the 40% of the full cost of the woodworking machine and divided by the hours of machine

Fig. 9.19 Influence of cutting speed and tooth bite on the specific cost of machining. Beech wood $H = 1$ mm, $\rho_{max} = 50$ μm

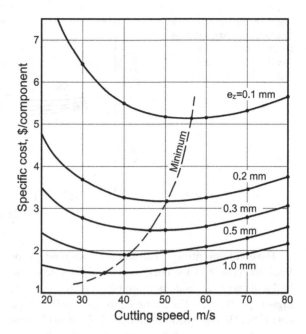

utilization per year ($/h). Generally, energy costs will be taken into account only for sanding.

Using Eq. (9.21), the tool life T reads

$$T = \left(\frac{\rho - \rho_0}{A}\right)^{1/m} \cdot \frac{0.0737}{v^{(n/m)+1} \cdot \sqrt{H/R}} = \frac{B}{v^{(n/m)+1} \cdot \sqrt{H/R}} \tag{9.27}$$

The unit cost of components is simply given by

$$C_u = \frac{C}{N} \tag{9.28}$$

The following example shows the use of these relationships (see also Eqs. (9.20) and (9.22)). Tool diameter $D = 120$ mm, $z = 4$, surface quality constraint $\rho_{max} = 50$ μm, $e_z = 0.5$ mm, $H = 1$ mm, $n = 6000$ rpm, $e = 12$ m/min, $L_f^c = 10$ m, the setup time $t_s = 10$ min, $t_{ch} = 5$ min, $t_n = 1$ min, specific machining cost $x = 50$ $/h = $0.833/min, tool sharpening cost $y = 10$ $. Using Eq. (9.27) and with $1/m = 1.92$ and $(n/m) + 1 = 2.25$, the tool life is 127.8 min. The machining time is calculated from Eq. (9.25) with $K = 1.1$ and it results $t_m = 0.916$ min. The number of components produced during the life of the tool is $N = 127.8/(0.916 + 1) = 66.7$. The total cost per time cycle is $129 and the unit cost is $C_u = 129/66.7 = $1.935/piece. This result may not be optimum. We easily can find the optimum by varying the cutting speed and the tooth bite. These calculation results are illustrated in Fig. 9.19 (Csanády et al. 2019).

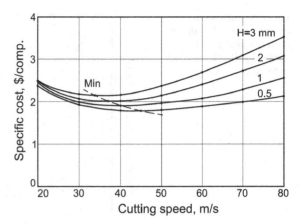

Fig. 9.20 Specific cost as a function of cutting speed for different depths of cut. Beech wood, $e_z = 0.5$ mm, $\rho_{max} = 50$ µm

The optimum cutting speed and its corresponding specific cost depend on the tooth bite. The course of the curves on the right side of the minimum line (dotted line) is flat allowing a good compromise to increase the production rate with a slightly higher specific cost. For example, using $e_z = 0.5$ mm and $H = 1$ mm, the minimum cost is \$1.93/component and the feed speed is 13 m/min. Selecting a 60 m/s cutting speed, the unit cost will be \$2.10 (9% higher) but the feed increases to 19.1 m/min which means a 46% increase in the production rate.

The depth of cut modifies the effect of cutting speed on the unit cost. Increasing the depth of cut decreases the tool life and increases the unit cost. The material removal rate (or feed speed) always surpasses the rate of cost increase when using a higher cutting speed, Fig. 9.20 (Csanády et al. 2019).

For cut 3 mm deep, the optimum cutting speed is 34 m/s but increasing the cutting speed to 60 m/s, the unit cost will increase 24% while the feed speed increases 76%. This allow to make compromise between minimum cost and maximum production rate.

To optimize sanding, the main governing factors are the amount of material to be removed, the cutting speed, feed speed, and surface pressure. The constraints are the required surface roughness and the service life of the sandpaper. There are differences between aluminium oxide and silicon carbide abrasive grains. Silicon carbide abrasives due to their higher heat conduction coefficient are better for sanding composite materials containing adhesives. Generally aluminium oxide abrasives perform better for sanding solid wood.

Using the same nominal grit size, the grit geometry of aluminium oxide and silicon carbide may slightly differ. Grits of aluminium oxide are more rounded with a slightly higher equivalent radius. Aluminium oxide abrasives gave somewhat higher average roughness R_a compared to silicon carbide (de Moura and Hernandez 2006).

Furthermore, the grits of aluminium oxide produced deeper surface damage, similarly to a dull knife (de Moura and Hernandez 2006; Ratnasingan and Scholz 2004).

The cost of *sanding* is related to the feed length or the sanded surface in the unit time, and occasionally to the material removal rate. The cost elements are similar to those of knife machining. The machine and labour costs can be given by

$$C_m = \frac{1}{v_f}(C_{ma} + C_{lab}) \cdot \left(1 + \frac{t_n}{t_{pr}}\right) \text{ \$/m} \tag{9.29}$$

where

v_f is the feed speed,
C_{ma}, C_{lab} are the machine and labour costs in the unit time,
t_n, t_{pr} are the non-productive and productive time.

Cost of the sanding belt is related to its service life T:

$$C_b = \frac{1}{v_f} \cdot \frac{C_{belt}}{T} \text{ \$/m} \tag{9.30}$$

The cost for changing the sanding belt depends on the tool life and time needed to change it (t_{ch}):

$$C_{ch} = \frac{1}{v_f} \cdot \frac{t_{ch}}{T}(C_{ma} + C_{lab}) \text{ \$/m} \tag{9.31}$$

Sanding machines consume much energy and therefore, their energy consumption must be taken into account. A simplified calculation (see in Sect. 4.7) gives the following relationship

$$P = 0.95 \cdot p \cdot A \cdot v_c \text{ W} \tag{9.32}$$

where

p is the surface (platen) pressure,
A is the sanded surface, $A = b \cdot L_c$,
v_c is the cutting speed.

Taking one hour productive time and the unit price of electricity C_{el} (\$/kWh), the specific cost of energy is

$$C_e = \frac{P \cdot C_{el}}{60 \cdot v_f}\left(1 + \frac{t_n}{t_{pr}}\right) \text{ \$/m} \tag{9.33}$$

where P must be substituted in kW. The main problem is the correct determination of the belt life. The most reliable method is to determine the economic lifetime based on the material removal rate. The following example will demonstrate this evaluation

Table 9.1 Material removal in the subsequent 10 min intervals and the corresponding specific costs (calculated from Dobrindt 1991)

Intervals (min)	0–10	10–20	20–30	30–40	40–50	50–60	60–70	70–80
$Q.10^{-3}$ cm^3/min	12	7.26	5.05	3.47	2.526	1.89	1.42	1.1
$/m^3	4166	2855	2468	2340	2310	2330	2360	2448
$/m^2	2.08	1.146	0.83	0.677	0.58	0.52	0.476	0.44

method. Taking a belt sander 1 m wide, 30 cm sanding length, surface pressure $p = 0.5$ N/cm^2, belt speed 20 m/s, $v_f = 8$ m/min, to sand beech wood (Dobrindt 1991). The sanded surface is 0.3 m^2, the power requirement is approximately 28.5 kW. The maximum removal rate is 0.45 cm^3/cm^2 min, which means 1350 cm^3/min for the entire sanded surface. This maximum value continuously decreases due to the belt wear. Using the experimentally obtained removal rate as a function of time, for the next ten minute intervals the material removal can be calculated as shown in Table 9.1.

Taking the machine cost $20/h, labour cost $10/h, sanding belt $20 and setup time of 20 min, time for belt change 30 min, the operational time is variable. If the calculated cost at the end of each interval is divided by the volume of removed material then we obtain a minimum value at 50 min (Fig. 9.21). At the same time, the cost related to the sanded surface area is continuously decreases. The belt should be used as long as the surface quality requirements are met. However, if the number of passes for a given workpiece, due to the wear process, increases then the finished surface area in the unit time decreases and the cost function has an optimum as a function of working time around 50–60 min. Sadly, no well-established quality criterion has been worked out yet.

The continuous decrease of specific stock removal due to wear may cause some problems in the sanding operation. If the thickness of material removal is important then it may be corrected by varying the surface pressure or feed speed. Fortunately, a

Fig. 9.21 Optimum lifetime of a sanding belt related to stock removal and the specific cost of sanding a surface area

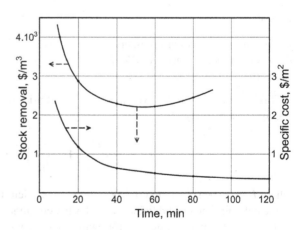

tenth of a mm deviation is generally not a problem in evaluating the size accuracy. If a given thickness of the workpiece is required, a contact wheel sander will be used.

9.5 Machining Accuracy

Design of tolerances in a new product development should be in accordance with the machining accuracy of parts. If it is not the case, either machining capabilities will not be utilized or unacceptable (rejected) parts due to machining errors will be high (Csanády et al. 2019). That means that machining accuracy, tolerance and number of rejected parts are closely correlated.

Concerning machining accuracy, we may distinguish between two cases:

- variation in dimensions shows a stochastic character with a given standard deviation, but the mean value corresponds to the design value ($\Delta\mu = 0$),
- systematic machining error occurs (shifting of the design value, $\pm\Delta\mu$) wich is manly due to setting errors.

The accuracy of machining depends on many influencing factors:

- machine rigidity and damping,
- accuracy of spindle running, bearings,
- clamping of workpiece,
- sharpness of tool,
- running circle accuracy,
- tooth bite,
- feed speed,
- relative workpiece mass (g/cm) in through-feed machines,
- spring constant of press rolls,
- mechanical properties and their uniformity of the processed material,
- resolution of setting mechanism,
- skill and experience of workers.

These factors contribute to the stochastic error, except for the last two which manifest itself in systematic error.

Variation of the machined dimensions with respect to the mean follows normal distribution when a number of causes are acting in a random way, each with small effect as related to the total variation.

The type of tolerance, which may be symmetric asymmetric and one-sided, has also a considerable effect on the interaction of tolerance and machining accuracy. Figure 9.22 (Csanády et al. 2019). Asymmetric tolerance and shifting the design value increase the number of rejected parts if the tolerance width remains unchanged (see in Fig. 9.22).

In order to characterize the interaction of tolerance and machining accuracy, a characteristic number C is defined which is the ratio of tolerance width ΔT and the threefold of standard deviation.

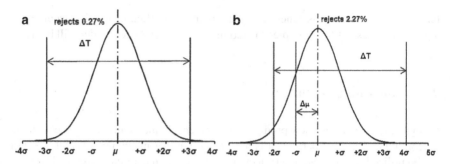

Fig. 9.22 Symmetric tolerance (**a**) and asymmetric tolerance with shifting of the design value (**b**) and their interaction with machining accuracy

For symmetric tolerance and without shifting of the mean value ($\Delta\mu = 0$) this ratio is given by

$$C = \frac{\Delta T}{3 \cdot \sigma} \tag{9.34}$$

In general case: with asymmetric tolerance and shifting the design value, the above equation reads:

$$C = \frac{\Delta T \mp \Delta T_{as} \pm \Delta\mu}{3 \cdot \sigma} \tag{9.35}$$

where ΔT_{as} is the shifting of the tolerance limit in relation to the symmetric case ($\Delta T_{as} \approx -\Delta T/3$ in Fig. 9.22b).

If the tolerance width ΔT equals the $\pm 3\sigma$ range then $C = 2$ and the ratio of unacceptable parts (rejects) on the two sides is 0.27% (see Fig. 9.23). If the value of C decreases, the number of rejects increases. Asymmetric tolerance and shifting of the design value mean a worse case compared to the symmetrical case with no shifting ($\Delta\mu = 0$).

The asymmetric tolerance can be compensated easily however, setting an accurate shifting $\Delta\mu$. If $\Delta T_{as} = \Delta\mu$, then we return to the more favourable symmetric case of Eq. (9.34).

Figure 9.23 demonstrates that the numbers of rejected parts sharply increases if no accordance between tolerance and machine capability.

Experimental measurements results of component manufacture on a four-sided moulder is given in Fig. 9.24 (Kovács et al. 2011). 500 mm long workpieces of three hardwood species (beech, oak and black locust) 60 of each were machined to width and thickness, given in Table 9.2. Thickness and width of the machined workpieces was measured and the mean and standard deviation were calculated. Due to the varying densities and sizes, the mass of workpieces related to the unit width (g/cm) was also different. Latter may influence the vibration behaviour of workpieces in trough-feed planers and hereby also the machining accuracy (see in Sect. 5.5).

Fig. 9.23 Expected number of rejected parts as a function of relative machine capability coefficient C

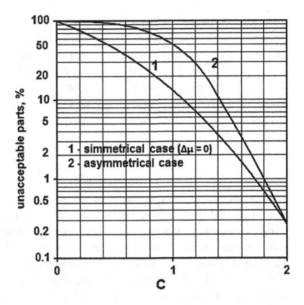

Fig. 9.24 Measured standard deviation of workpieces as a function of final width and calculated 2σ and 3σ value for tolerance design, and the corresponding probability levels

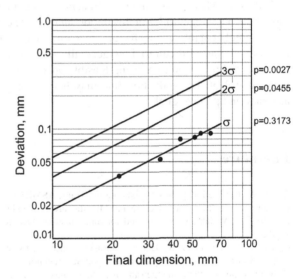

The applicable tolerance corresponds to dimensional variations at 3σ which ensures an acceptable number of rejects of 0.27%. Taking smaller tolerances, the percent of rejects strongly increases.

The vibration amplitude of workpiece in a through-feed machine is inversely related to the mass of the workpiece (see Sect. 5.5). In the above described experiments the mass related to the unit width fluctuates between 100 and 250 g/cm which affects the machining accuracy as a function of width.

Table 9.2 Nominal cross-section dimensions of work pieces used in process capability study

Species	European beech		English oak		Black locust	
Dimension	Width (mm)	Thickness (mm)	Width (mm)	Thickness (mm)	Width (mm)	Thickness (mm)
Raw	57.0	48.0	68.0	38.0	61.0	27.0
Final	51.5	44.0	62.5	34.5	55.5	21.5
SD mm	0.084	0.078	0.091	0.0517	0.088	0.037
Density g/cm^3	0.7	0.7	0.71	0.71	0.8	0.8

In an other series of experiments the machining accuracy of tenons was measured on a single end tenoning unit of a window machining centre (Kovács et al. 2011). Window frame parts, made of pine wood, with tenons of 8, 12 and 30 mm thick were used. The standard deviation σ was varied between 0.05 and 0.08 mm depending on the tenon thickness. In this case a less tenon thickness means less efficient support of the machined end of workpiece giving slightly higher machining error. But the machining errors have the same magnitude as in Fig. 9.24.

Sadly there are few experiments on machining accuracy in the literature and, therefore, the general regularities of this important machine characteristics are not known. A great number of influencing factors have some kind of effect on the machining accuracy, which make difficult to draw generally valid concludes. Keeping the machine and tools in good condition, using feasible operational parameters, making continuous control of the production line may help to preserve the desired machining accuracy and product quality.

Literature

Carter, W., Ragsdell, K.: The optimal column. Trans. ASME 71–76 (1974)

Csanády, E., et al.: Optimum Design and Manufacture of Wood Products. Springer, Berlin (2019)

Davidon, W.: Variable metric method for minimization. ARC Res. and Dev. Report, ANL-5990 (1959)

de Moura, L., Hernandez, R.: Effect of abrasive mineral grit size and feed speed on the quality of sanded surfaces of sugar maple wood. Wood Sci. Technol. 517–530 (2006)

Déry, J.: Spanungsuntersuchungen an einer Blockbandsäge. Act Fac. Lign. Sopron, S. 43–52 (1985)

Dobrindt, P.: Optimiertes Schleifen von MDF- und Spanplatten. Workshop Tagungsband, 8.Holztechn. Koll. Braunschweig, pp. S. 125–136 (1991)

Feld, H., et al.: Carbide selection for woodworking tooling. In: Proceedings of 17th IWMS, Rosenheim, pp. 520–527 (2005)

Fletcher, R., Powell, M.: A rapidly convergent descent method for minimization. Comput. J. 163–168 (1963)

Garcia, I.: New developments in ultrafine hardmetals. In: Proceedings 17th IWMS, Rosenheim, pp. 534–542 (2005)

Kovács, Z., et al.: Investigation of the possibilities of enhancing the effectiveness of furniture part manufacture. Res. Report, University of Sopron (2011)

Licher, E.: Schnittwert-Datenbank für die Holzbearbeitung. 8. Holztechn. Koll. Braunschweig, S. 33–58 (1991)

Ratnasingam, J., Scholz, F.: Wood Sanding Process. University Putra Malaysia (2004)

Saljé, E., Drückhammer, J.: Qualitätskontrolle bei der Kantenbearbeitung. Holz Roh Werkst. S. 187–192 (1984)

Saljé, E., Drückhammer, J.: Kantenschartigkeitmessengen an beschichteten Spanplatten. Holz-Zbl. S. 1782–1784 (1985)

Sitkei, G., et al.: Schnittleistung und Energiebedarf von Gattersägen. Acta Fac. Lign. Sopron, S. 23–30 (1988)

Sitkei, G., et al.: Theorie des Spanens von Holz. Fortschnrittberichte No. 1, Acta Fac. Lign. Sopron, S. 72 (1990)

Stadler, W.: Multicriteria optimization in engineering and sciences. Plenum Press, New York (1988)

Timoshenko, S., Goodier, J.: Theory of Elasticity. New York (1951)

Index

Printed in the United States
by Baker & Taylor Publisher Services

Printed in the United States
by Baker & Taylor Publisher Services